Social Responses to Technological Change

Recent Titles in Contributions in Sociology
Series Editor: Don Martindale

Max Weber's Political Sociology: A Pessimistic Vision of a Rationalized World
Ronald M. Glassman and Vatro Murvar, editors

The Adventure of Reason: The Uses of Philosophy in Sociology
H. P. Rickman

Social Hierarchies: Essays Toward a Sociophysiological Perspective
Patricia R. Barchas, editor

Social Cohesion: Essays Toward a Sociophysiological Perspective
Patricia R. Barchas and Sally P. Mendoza, editors

Family and Work: Comparative Convergences
Merlin B. Brinkerhoff, editor

No Place to Hide: Crisis and Future of American Habitats
Manuel Marti, Jr.

Systems of Discourse: Structures and Semiotics in the Social Sciences
George V. Zito

Ideologies, Goals, and Values
Feliks Gross

People's Poland: Patterns of Social Inequality and Conflict
Wladyslaw Majkowski

Weeding Out the Target Population: The Law of Accountability in a Manpower Program
James Latimore

Social Responses to Technological Change

EDITED BY

Augustine Brannigan

AND

Sheldon Goldenberg

Contributions in Sociology, Number 56

Greenwood Press
Westport, Connecticut • London, England

The editors gratefully acknowledge the support given this colloquium by the Social Science and Humanities Research Council of Canada, the University of Calgary and the Department of Sociology, the University of Calgary.

Library of Congress Cataloging in Publication Data

Main entry under title:

Social responses to technological change.

(Contributions in sociology, ISSN 0084-9278 ; no. 56)
Bibliography: p.
Includes index.
1. Environmental policy—Addresses, essays, lectures.
2. Technological innovations—Social aspects—Addresses, essays, lectures. I. Brannigan, Augustine, 1949-
II. Goldenberg, Sheldon. III. Series.
HC79.E5S58 1985 363.7′052 84-27934
ISBN 0-313-24727-7 (lib. bdg.)

Library of Congress Catalog Card Number: 84-27934
ISBN: 0-313-24727-7
ISSN: 0084-9278

First published in 1985

Greenwood Press
A division of Congressional Information Service, Inc.
88 Post Road West
Westport, Connecticut 06881

Printed in the United States of America

The paper used in this book complies with the Permanent Paper Standard issued by the National Information Standards Organization (Z39. 48-1984).

10 9 8 7 6 5 4 3 2 1

Copyright Acknowledgments

Portions of Allan Mazur's "The Mass Media in Environmental Controversies" are adapted from *The Dynamics of Technical Controversy* by Allan Mazur (Washington, D.C.: Communications Press, 1981).

Allan Schnaiberg's "Did You Ever Meet a Payroll? Contradictions in the Structure of the Appropriate Technology Movement" previously appeared in the *Humboldt Journal of Social Relations* 9, no. 2 (Spring/Summer 1982): 38-62, and is reprinted with permission of the *Humboldt Journal of Social Relations*.

Stephen G. Peitchinis's "Microelectronics, The Concept of Work and Employment" is taken from *Computer Technology and Employment* (London: Macmillan Press Ltd., 1983 and New York: St. Martins Press, 1983) and is reprinted courtesy of Macmillan Press and St. Martins Press.

This book is respectfully dedicated to all of the contributors who made it so successful a colloquium, and whose continued intense concern is clearly evident in this collection of papers.

Contents

Exhibits

I
Introduction

1

Technology, Society, and Change: Problems and Responses

AUGUSTINE BRANNIGAN
and
SHELDON GOLDENBERG

TECHNOLOGY, HARMFULNESS, AND THE RISE OF THE ECOLOGICAL CONSCIOUSNESS

Throughout history there has been an optimism associated with the rise of science and its capacity to ameliorate the human condition. This feeling reached a high point following the Second World War with the rise of the consumer society and the creation of material wealth and prosperity on a scale seldom anticipated in earlier history. The consumer society also brought with it a host of social programs financed by taxes on the new wealth, waves of immigration and urban growth, as well as greater political emancipation for minority groups and women. It was in this context of unrestrained growth and prosperity that the first warnings about the unintended consequences of the technological society emerged. The initial warnings of Rachel Carson regarding environmental pollution and Paul Ehrlich's fears of the population explosion were joined increasingly over the 1960s and 1970s by numerous others expressing fears over a panoply of disasters from acid rain to nuclear meltdowns to carcinogenic food additives.

Social critics have coined the term "technophobia" to describe the nagging misgivings and distrust felt toward the scientific transformation of social life. The 1960s saw the resurgence of the equity-oriented environmental movement which sought, often rather naively, the "re-greening" of America. Within a decade this had been transformed into a concern for technological impact assessment, as governments produced institutional attempts (the Environmental Protection Agency, Environmental Impact Assessment, Social Impact Assessment) to translate some of these "concerns" into practical decision-making rules. In the 1980s we see a bimodal response, ambivalent, hesitant, and not a little fearful. There appears to be a resurgence of concern for all things nuclear, whether reactors or bombs, as well as a concern for the effect of science and technology on the structure of employment, particularly with the introduction of computer technologies and the booming unemployment crises that are just around the corner. Equity concerns and collective rationalism are again a focal point of the opposition. But we also find the reaction: "yes—environmental sensitivity, but not at the price of large-scale unemployment!" In other words, we find a willingness to forego equity issues if unemployment is the price to be paid for pollution control, or if a choice is forced between jobs and a cleaner, safer, fairer environment. The revived ecological consciousness of the 1960s has reappeared in a new guise in the 1980s, based now on the recognition that the mastery of nature through science and technology involves a sort of Faustian bargain with the Devil: the material benefits of the technological society must be paid for, usually unwittingly, in terms that involve the possible destruction of, or risks to, health, life, and the environment.

SOCIAL RESPONSES TO TECHNOLOGY

This volume was prepared following a conference on the social responses to technological change and the environmental crisis.* The point of the conference was *not* to document the ecological evidence of harm or describe the most recent advances in biotechnology or microcomputers and their probable consequences. Rather, the point was to examine the social reactions to the changes produced by technological modernization. If modernization has been accompanied by an undercurrent of dangerous or lethal harm, what social forces or social processes has this mobilized?

One of the basic issues implicit in our title, and explicit in many of the papers, concerns the relationship of public policy to technological change. Does technology determine social change or is technological

*The conference was supported with funds from the Social Sciences and Humanities Research Council of Canada and the University of Calgary, March 24, 25, 26, 1982.

modernization mediated by public policy? Must society adapt to technological development as best it can and "pick up the pieces" after the fact, or can public policy direct, restrain, and control technological innovations and impact?

In the papers that follow, we have identified several major, though not always discrete, areas of social response to the dynamic factors that have affected society's relationship to nature.

Social movements. Given the pervasiveness of technological and environmental woes confronting modern life, it is hardly surprising that numerous social movements have been generated. Grass-roots public protests, reflected in the environmental movements as well as the peace movements of the 1960s and the 1980s, constitute the most obvious social responses. However, as Allan Schnaiberg shows in the first paper of this volume, "the environmental movement" is really a misnomer since it has enjoyed neither a consistent constituency nor a set of fixed political objectives across time. According to Schnaiberg, it has evolved from a concern with the social questions of the unequal distribution of wealth and power to a more technical concern with the format of production. In a second paper, Schnaiberg returns to the contradictions which the movement's recent concern for technical production-related issues has created. This new focus bodes ill for the employment prospects of the many disadvantaged groups in society for whom the earlier movement had sought more equal opportunity and treatment. This departure from equity goals may explain in part the impotency of environmental social movements to effectively set political agendas aimed at the achievement of desired social change. This is reflected in the current acid rain dilemma: productivity spells pollution, and pollution control spells unemployment, a point at the heart of Andrew Thompson's paper in the section on legal controls on the negotiation of control over and sanctions attached to the negative environmental impacts of the West Coast fisheries. Similar contradictions between labor and productivity are addressed by W. G. Carson and Stephen G. Peitchinis in later sections of this volume.

Legal controls. If a neighbor poisons another neighbor's well, one would presume that a criminal charge could be laid for attempted murder or willful damage to property. However, if the first neighbor happens to be a mining, smelting, or forestry company and the poisoning of the ground water is an unintended by-product of an industrial process, where does the aggrieved party stand? Surely the victim has some protection under the law and can seek a criminal, civil, or administrative solution. Surprisingly the control of technological harm by law is both ambiguous and relatively ineffective. Criminal remedies are usually unavailable because of the absence of the motive to harm; no "guilty mind" means no criminal conviction. Consequently victims of Three Mile Island, Love Canal, Agent Orange, sour gas emissions, food additives, faulty consumer products, and unsafe working conditions have no criminal law remedy. An individual may take a private action. However, civil remedies must be initiated by aggrieved parties directly, are expensive and time-consuming, and put the onus on the victim to show the harm. As for the last option, government monitoring of industry is done via administrative law. Both papers in this section describe a situation of bargaining between the state and commercial actors to determine the nature, extent, and conditions of government regulations. In other words, the public must "bargain" for the safety and health of workers and the environment. However, the authors here trace the origin of this loose control to rather different roots. Andrew Thompson argues that the bargaining between state and industry over the enforcement of environmental pollution standards arises from the scientific and legal uncertainties associated with detecting harm, as well as litigating cases. W. G. Carson, in a study of the ineffectiveness of safety regulations in the offshore petroleum industry, traces the looseness of control to political factors: chiefly to governmental desire to speed the rate of development in order to vitalize the economy, whatever the adverse effects in terms of worker injuries and deaths. Presumably in Carson's view, most of the harmful and unintended damages of the industrial processes could be traced to the state's fear of restricting or limiting the freedom of "free enterprise," despite the cost to the public. As a result, in both papers, the efficacy of legal controls is disputed, but for vastly different reasons. The conflicting explanations have altogether different policy implications.

The media. Another area examined is the role of the media in amplifying and/or containing public reactions to environmental impact. This involves not merely the role of the news media in the formulation of public opinions concerning issues like nuclear power, but also the role of the entertainment media, including television and movie dramas. Their effect may well be to stereotype and reinforce certain images, often hysterical ones, in stories that exploit the public's fear of scientific advances. Allan Mazur focuses on the reportorial role of the news media in the Three Mile Island coverage, while Dorothy Nelkin traces the public images of the atom as portrayed by Hollywood in the movies.

Public policy, technology, and the control of change. As noted earlier, these collected papers do not deal primarily with technological developments, or even with their purely environmental impact. They are concerned first and foremost with the social responses to technology. Questions are raised concerning the relationship between public policy and social structure, and between social structure and technological impact. In the various cases examined in this volume, consideration is given to the question of causality and causal ordering. What effects do technology, social structure, and ideology have on one another? Does technology affect social structures directly or is this effect mediated by public policy? Does social structure affect technology or ideology directly? In every instance the social structural variable must be carefully examined. This is not a new type of puzzle for social scientists to examine.

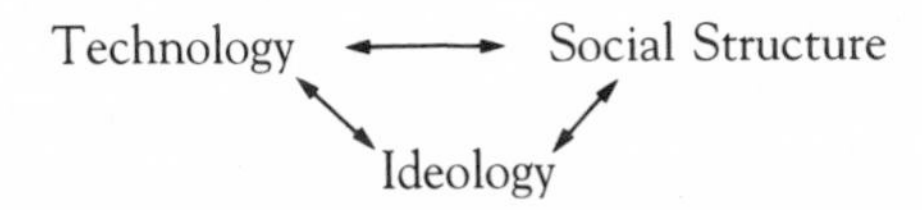

The three axes or variables are recurrent in these papers. And a great deal of social science literature has to do with the relationships depicted here in such broad terms. How closely are these factors interwoven? How much autonomy has any single one?

One of the clearest statements of these relationships is critically explored by William Leiss in his paper on the information society. The original thesis about the information society held that microelectronic technology would automatically revolutionize society. Policy was thought to be limited to whether or not governments ought to facilitate or resist the inevitable transformation of our society. As Leiss shows, recent pronouncements from Ottawa and Washington have precisely this character. All urge us to "get on with it" before we are swallowed up by technological changes beyond our control. In other words, technology is believed to impact deterministically on social structure, while public policy is to respond to such impacts, whether by facilitating or by striving to oppose or to buffer them in some way. Leiss contends that policy may have more independence than this and that policy decisions should not be forced by mere visions of what is an indeterminate future, still malleable and open to many possibilities. This view suggests that public policy can strongly affect technology rather than being restricted to a simple passive role. Policy choices can affect social conditions at least, thereby making some technological effects more viable and others less so. Policy can also encourage or discourage technology directly, and we must make this choice actively and not default on it by assuming in advance that technological impacts will occur in any event. Such default also constitutes a policy and has effects.

Stephen Peitchinis makes a forceful case for technological effects as paramount. He feels that we must come to grips with the implications of "jobless growth" created by the wholesale advent of the microprocessor. For him, there are fairly clear social structural implications, and these in turn require policy shifts. Technology is the key, and policy responds after the fact. One is strongly reminded of William F. Ogburn's theory of culture lag in these formulations (1922). Like Ogburn, Peitchinis suggests that policy and structure only slowly adapt to changes forced on them unwittingly by technology. Andrew Thompson, in a discussion of the pollution of fishery stocks, suggests a legal policy response to technology that is sensitive to the powerful effects of that technology and tuned finely to its changes. The determinant or key variable is technological: mining and forestry produce calculable harm to the environment. Policy here is a vehicle for institutionalizing new patterns of response to the changes in social relations and social structure that result, just as in Ogburn the advent of the automobile had far-reaching effects on the social system and required a new set of laws, institutions, and policies in order to manage these effects in a stable and orderly manner. Dorothy Nelkin's paper suggests that representations of social change in the entertainment media clearly identify technology as the determining factor, even when such innovations appear to be out of control and threatening human survival.

Not all our contributors agree that technology determines or produces social change. Various commentators point to the social relationships in society that control how and where technologies get used, and the uneven distribution of the costs and benefits of modernization. The analysis of elites and interest groups becomes central here. For example, W. G. Carson lays the responsibility for occupational safety and hazard in the North Atlantic, not on the perils of the technology involved (the ideology of technological determinism), but on vested interests in the state and oil companies, interests that were (are?) better served by ignoring or failing to implement the kind of strict provisions that would ensure greater safety but that might have delayed development. Similarly, the two papers by Schnaiberg describe changes over time in the "movement" response to technology—from the earlier period in which conservation and equity were primary concerns to the current period in which the format of production has become primary. Schnaiberg suggests that such changes in ideology and its associated social movement structure are not simply the effects of technological change, but must be analyzed in a considerably deeper manner, relating vested interests to ideology and ideology to public policy, and these to the encouragement and format of technology.

In this dialectic between the imperative of technology and the question of social dominance, of what relevance is *public policy*? From the first perspective described—the technological—policy is adaptive and reactive. It attempts to bring society to a new state of equilibrium that is technologically wrought. Policy has no causal weight itself. It is epiphenomenal or superstructural and ideological or expressive only. From the second perspective, policy responds to and embodies the vested interests of various parties in society, each in pursuit of its own gain. "Public" policy is perhaps a misnomer, since the most vulnerable party to the contemporary effects of technological society is the "public" at large, reflecting the historical "tragedy of the commons." From this perspective—the structural—policy is a conditional variable that alters technology by encouraging some aspects and discouraging others, by distributing costs and benefits according to a political agenda. From the equity theme, one can discern a third possible relationship of policy to social structure and technology. From this perspective, public policy has its own important effects on both of these other aspects of the model. In a cautious and somewhat stereotypical "liberal social scientist" view, it is hoped that public policy will be efficacious and equitable, not merely constitute public relations for vested interests, but a passionate attempt to govern, legislate, and control technology and society for the "social good."

SOME CHARACTERISTICS OF THE SOCIAL RESPONSES

The papers in this collection cover a large number of substantive areas. Nonetheless, there are recurrent common themes, which revolve around several ambivalent or contradictory tensions.

Equity versus efficiency. In responding to the costs of pollution or industrial risks, which value, equity or efficiency, takes priority? Schnaiberg raises the question in terms of the phases of development in the environmental movement: the initial preoccupation with the exploitation of natural resources by certain groups, whether as countries or classes, at the expense of others is superseded by the later focus on soft energy paths and Appropriate Technology (A-T). This shift has transformed the "social" problem into a "technical" one: does industry need to be as large or as capital-intensive as current trends suggest? Will alternative, smaller scale developments not suffice? Schnaiberg's view is that the A-T movement has transmuted the redistributional equity goals of the early movement into merely technical organizational issues that in some ways contradict and in many ways ignore the original equity concerns. In his view, the environmental movement has appeared in different guises at different times, supported for different reasons with different intensity by different groups. It appears that equity concerns increase only when the economy booms and fall as self-interest rises in periods of growing unemployment. It is as though the pursuit of collective rationality can attract sympathetic activists on a large scale only in times of economic prosperity.

The equity and efficiency question also emerges in the papers by Carson and Peitchinis. Carson points to the unequal distribution of the costs and benefits of rapid petroleum development: in the cost-benefit equation, the costs in terms of accidents and deaths were allocated to the working classes, while the benefits went disproportionately to the multinationals and their investors. In his comment on Peitchinis, Donald L. Mills likewise stresses the unequal marginalization of workers in blue collar professions with the introduction of microelectronic technology. While making many jobs more efficient, the computer revolution strikes disproportionately at blue collar and clerical workers who are least organized, most vulnerable, and often female. To his credit, Peitchinis deals primarily with these equity issues, and his thesis in large measure proposes an optimistic solution to the economic problems introduced by technology that could otherwise have such dire and inequitable results. In sum, the papers frequently return to the dilemma of what is sound in economic terms versus what is desirable in terms of social equality.

Rationalism and self-interest. The fact that the costs and benefits of technology are differentially distributed to different social classes and groups points to another opposition—the rationality of technology for the whole society versus the self-interest of the parts. Technologies do not affect societies "as a whole" except in the most abstract sense. They are drawn into social relations either as commodities (home computers) or as part of the process to produce such commodities (nuclear power in manufacturing). Consequently the extent of effects—both good and bad—on various groups reflects what Marx called the "relations" of production. Those who are able to control the productive process are able to "capitalize" on innovations that pay dividends and/or distribute the liabilities of technology such as pollution to others. In terms of immediate self-interest, such conduct is rational, even though for the society as a whole the unequal distribution of costs and risks is irrational. For instance, speaking economically, the costs of acid rain (versus pollution abatement equipment) might be best passed to the public sector by individual entrepreneurs; yet the cost to the collectivity in terms of the resultant destruction of the commons would be beyond measurement. Rational self-interest of the parts, contrary to conservative economics, is irrational for the society as a whole. Herbert Marcuse drew the same conclusion in *One Dimensional Man* when he spoke of the irrationality of the contradiction of forces marking the modern societies. His 1961 remark that nuclear weapons were the very source of the fear for which they were required is as apropos today, as Europe becomes the staging ground of new SS20 Cruise and Pershing missiles. Our missiles are required as a defense against their missiles and vice versa. Individual rationalism produces a fearsome insanity of the whole. No mutual agreement has overcome the antagonism of the parts. No social contract has unified the conflicting interests.

The ideology of technological determinism. William Leiss speaks of the role of ideology in the "construction" of the whole concept of technological determinism, and Allan Mazur documents the power of the press in the creation or negotiation of the meaning of technological innovation and impact. William Reeves's comments on technological determinism, and those by J. Richard Ponting on the role of the media in technical debates, also point to the conditional nature of technological impact and the manner in which ideological and structural factors influence society. Clearly the idea that technology controls social change is a social construction, if not an outright ideology. This theme is reflected in Schnaiberg's papers, in terms of both the historical dimension as conservation has changed its meaning and implications over time, and in terms of A-T and its promotion by adherents. From this perspective, the impact of technology and its character, its direction and implications, though touted as inevitable, are continually negotiated. Automation *may* produce jobless growth and a post-industrial era of plenty. Many ideologists and "futurists" have assured us that this is so and that, while there will be temporary dislocations and some differential impacts, the long-term result is clear and policy ought to encourage such development in every way. The debates in this volume suggest otherwise. For example, Donald L. Mills argues that automation and the advent of the microprocessor are unlikely to produce a utopia in either the near or distant future. The dislocations and disruptions are obvious and quite unequally distributed, in ways that return our attention to issues of equity, of social mobility, of stratification and conflict. These in turn recall the question of the role of policy in controlling change.

Public policy, voluntarism, and sociological imperatives. One of the chief dilemmas of the social sciences is the question of whether human conduct is voluntarist or socially determined. Voluntarists stress free will and the role of wishes, plans, and intentions in explaining behavior. Social determinists, whether of a behavioristic or economic bent, ignore subjective factors in favor of the objective conditions of conduct. The parallel in these papers is the conflict between a policy-controlled social change or structurally necessitated change. The structurally necessitated change is not equivalent to technological determinism. As noted above, a number of papers and their discussions are careful to deny that social change *inevitably* follows from advances in or from the effects of technological innovations. On the contrary, they seem to point to social forces that underlie the control and introduction of technologies. Carson's study of offshore petroleum installations and Peitchinis's discussion of microelectronic processors suggest that market forces have brought about the negative impacts of the respective technologies. Schnaiberg seems to imply that market forces will impede the development of A-T. Leiss stresses that market forces (and not the information "format" of society) will continue to form the elements of class society. In addition, for Thompson, the reality of the company's economic situation plays a part in producing a "bargaining" mode in the government's control of pollution. However, nearly all imply that these things *might* have been different had different public policies been followed. In other words, the course of technological impact could have been mediated by plans and intentions embodied in public policies. Hence, the offshore installation could have been safer; microelectronic processors could have been more rationally introduced as could the Information Society.

These advocates of "technological voluntarism" exhibit what may be a misplaced optimism in public policy. Robert Park, the early American sociologist, once commented that in America when we pass laws, "we might as well get up and dance for all they do is serve an expressive function." He was no doubt contemplating the fate of Prohibition. Laws against selling liquor did little more than express a displeasure with insobriety: they did not fundamentally redirect behavior. According to this interpretation, public policy-making is symbolic behavior: the controls it initiates are swamped by other factors, chiefly the economic. Furthermore, public policy might not even constitute symbolically the desired direction of change. It may be the *effect* of the change. Hence, it could be argued that economic realities produce the policies associated with them. For example, it seems reasonable to assume that the market determined the politics of Britain's exploitation of petroleum resources. The market also seems to promote the "bargaining" mode embodied in public policies. Put bluntly, public policy is often the effect of the change it is alleged to have caused.

This interpretation suggests that the social movement response is akin to Park's observation about people getting up and dancing. The behavior is expressive, not instrumental: it protests but cannot control. The legal responses given their recurrent ineffectual makeup similarly reflect his idea of policy: laws do not control development in a meaningful way; they more often remove the impediments for development.

The simple reductionism of such an interpretation overlooks a pervasive optimism found in this volume—an optimism that favors the collective advantages of public policies. How can the ideological role of desired policies be squared with the notion that policies either do not affect, and may even legitimize, a particular form of social change? Clearly the policy initiatives presented here do not legitimize the undesirable aspects of modern technological change. What they represent are alternative ideals and aspirations, alternative policies to those that have allowed the worst instances of social impact. It would be very naive to leave off our inquiry with the truism that there ought to be public policies to control or put a halt to the harm involved in technological change. This assumes that no policies have existed, while it is clear that in fact policies are already in place. However, these have not always or even frequently been devised with an eye to distributing costs and benefits equitably, effectively, or rationally. In view of the unequal distribution of social influence, this comes as no surprise.

Sociological determinism is not opposite to the optimism of a policy-controlled change. The determinism of sociology reflects the power of certain groups to *effectively* control the direction and content of economically driven change in spite of opposition. The effectiveness of the social interest groups lends a deterministic look to the sociological account that records this process. This suggests that it is not a matter of opposition between what people want versus what some reified society is structurally able to provide, but rather a question of which interest group in the society is able to affect public policy in reflection of its own interests. In weighing the various sources of social change—technology, rationality, social movements, the media, laws, and public policy—it is interest groups that strike sociologists as crucial. The dialogues between the contributed papers and their discussions highlight this point.

THE BASIC DILEMMA: INTEREST GROUPS AND THE LIMITS OF CONTROL

Sociological studies of the control of or responses to technological effects run hot and cold between the pessimistic assessment of problems and the optimistic invention of solutions. As indicated, our contributors frequently speak as though changes arising from technology are inevitable either because of technological determinism (that is, technology creates change) or technological voluntarism (technology emerges inexorably with the interests of market forces). Against this position there is a native sensibility in these papers which recommends that, despite how things are going, they could go otherwise. One detects in Schnaiberg, Carson, and Leiss a sympathy for equity. Similar sympathies arise in virtually all of the discussions. There is a sense that *public* policies vis-à-vis technology ought to be *social* policy; that is, that policy ought to reflect democratic interests. If our policies are in fact a reflection of the power of the vested interests only (i.e., if the industry controls its own "policing"), how can we change this so that policies are more truly social and broadly representative? How can public sympathy for equity be translated into social policy? How can the interests of an enlightened critic form a political agenda and mobilize a popular base of support? Few of our participants explicitly suggest guidelines for the conduct of social change, but several do provide implicit guidelines or suggestions. Leiss, Nelkin, and Mazur particularly emphasize the role of the media in creating and mobilizing our social consciousness. The media define conflicts and shape their course; a social policy, if it is to be a democratic expression, must encourage various publics to become more active users of the media and more active participants in the fight for people's minds. The restructuring of social relations so as to better embody social policy is the prime focus of the papers by Schnaiberg, Peitchinis, Carson, and Thompson. Here, too, are implicit guidelines for the achievement of social change.

The intellectual or social critic would appear to be important, if currently politically marginal, in the overall democratic process. The intellectual is important in contributing to an understanding of the problems and in discussing some possible solutions. However, turning this intellectual concern into a matter of public agenda may require more than the usual role of the media in researching technology news and defining areas as problematic. The identification of harm is clearly crucial here. This is rarely an easy task even if professional journalists pursue questions aggressively and freely. There must also be public interest and community responses of such magnitude as to revise existing political agendas so as to persuade the existing political structures to revise their policies.

At each stage in the development of public opinion, a loss or washout normally occurs. Intellectuals can obfuscate the issue. The media, especially where concentrated in ownership, can shape or misshape the public's concern or knowledge of problems. The community movements may be obstructed by apathy or public misinformation. And governments may "tough out" public protests, rendering them merely expressive. Despite these setbacks, public interest groups have nonetheless managed to affect agendas in such diverse policy areas as siting of the Cruise missiles in Europe and public sector layoffs in Canada. Public participation and participatory democracy are at work in industrial situations and in public situations of planning, however rare these instances and however meager their success. Public distrust of experts, and of the results of allowing them to create our world, appear to have grown in recent years into a broader movement, with practical applications in terms of such demands for decentralization and public participation. It is the long-shot possibility of similar responses to technology control that fosters sociological optimism in the face of what would appear to be social structural, as well as technological, near determinism. It is ironic, if not tragic, that the success of equity-oriented and truly public policy appears to depend on the success of the very system it seeks to alter. Only when Schnaiberg's "treadmill of production" is producing at a great rate is there the necessary luxury on which a movement to collective rationality and equity depends. Harnessing technological impacts means restricting and shaping them, and it appears that the will to do so is found widely only when such controls do not cost potential adherents their jobs or reduce their own standards of living. This is one of the key contradictions to which the following contributions are dedicated.

NOTE ON PRESENTATION

The papers presented here were subject to lively discussion and debate at the original conference. They were intended to provoke, and provoke they did. The papers oscillate between the idealistic and speculative on the one hand, and the grounded, the structural, and the Realpolitik on the other. Discussants were given free rein to criticize, confirm, and/or consolidate the key papers. In playing Devil's Advocates, the discussants debated many arguable assumptions in the original papers. In addition, they were able to amplify the basic theses in novel ways. Their inclusion in this volume adds immeasurably to the treatment of our topic—indeed, we could not justify omitting them. The discussants also establish the intellectual contradictions that remain at the heart of our subject matter. Our key speakers talk to key issues. Our discussants frequently take issue. In their infinite wisdom, the editors frequently found fault on both sides.

II
Social Movements

Professor Schnaiberg's introductory paper places the current version of an environmentalist movement in historical context. In doing so, he suggests that environmental concerns have been raised before and have been submerged before as well. Only a naive audience could believe that environmental progress is being steadily made or that, once recognized, environmental "problems" are readily solved.

From Schnaiberg's perspective, the recent rise of the Appropriate Technology (A-T) movement is to be viewed rather ambivalently at best, for it promotes a reorientation of the goals and strategies of the earlier environmental movement away from broad redistributive-political issues of equity and justice and toward more "technical" issues of cost-efficiency that do not challenge existing social relations.

2

The Retreat from Political to Technical Environmentalism

ALLAN SCHNAIBERG

THE DYNAMICS OF ENVIRONMENTAL MOVEMENTS: COMPETING PERSPECTIVES

The treatment of "the environmental problem" by contemporary social problems theorists and textbook authors is abundantly clear. Prior to the late 1960s, advanced industrial societies sustained dramatic economic expansion through the "free" use of natural resources. Most of the ecological results of this increasing usage of the natural environment went unnoticed until the attentions of authors like Rachel Carson (*Silent Spring*), Barry Commoner (*Science and Survival*), and Paul Ehrlich (*The Population Bomb*) roused a complacent public into a redefinition of this process as a social problem (Spector and Kitsuse, 1977). In turn, this led to the creation of a massive social movement—the environmental movement—which mobilized support for legislation and created a problem-solving set of agencies (especially the Environmental Protection Agency) and a concomitant set of legislative guidelines for protecting the environment. Legislation included the National Environmental Policy Act (NEPA) and Clean Air Act of 1969, and later legislation on endangered species, clean water, and, more recently, toxic substances. In recent years the "energy crisis" has displaced some of the concern of pollution abatement, and while the public has remained interested in protecting the environment, past successes have led to complacency and the movement base has shrunk considerably. Most recently there have been attempts to reduce environmental protection because of the economic recession and support for government deregulation of business—especially during the Reagan administration. Nevertheless, the movement has continued to fight for environmental protection, but at a reduced level, because of competing public concern for economic growth. Some support for environmental movement goals has come about through public and private attention to energy conservation.

With but a few variations, this is the essential social science tale of the environmental movement and its goals and means that is inculcated in undergraduate students. Ironically it differs in few essential ways from the tale told by the national news magazines in North America about the environmental problem. Indeed, one suspects that most authors of social problems texts have more frequently consulted these prolific media than the somewhat more limited literature on the environmental movement and the environmental problem. Depending on the author's predilections, the tale is either one of "sin" or "error." Our "sinful" version is that corporate greed dictated the polluting of our environment through harmful technologies—or even that we North Americans "have seen the enemy, and he is us," as Pogo put it many years ago. The "error" version is that our industrial society simply was inattentive to ecological processes until the 1960s and that we have struggled to belatedly correct this error of social intelligence (cf. Schnaiberg, 1980: chs. 6–7).

Neither the sin nor the error version of the cautionary tale squares with social and political history, however (cf. Schnaiberg, 1980: ch. 8; Albrecht and Mauss, 1975). As I have attempted to argue and document in my own recent work, we need a far deeper and more complex understanding of the modern "treadmill of production" to appreciate the roots of our environmental problems of excessive societal withdrawals from and additions to the natural biosphere (Schnaiberg, 1980: ch. 5). Moreover, we need to research the origins of modern environmental movements more cautiously, to appreciate the ways in which the environmental problem has surfaced historically in socially institutionalized ways. Such redefinitions of the environmental problem stem from the efforts of a variety of social movements that *can* be seen as an ongoing "environmental movement" at least throughout the twentieth century in the United States. Two facts stand out in the pre-1960 history of the American movements that raise doubts about both the "sin" and "error" versions of the social history of environmental problems.

First, there was substantial attention to ecosystem withdrawals and additions in the early part of the twentieth century, through the "conservation-efficiency" movement (Hays, 1969). Many contemporary problems of land and water use were delineated then, and professional movements arose, often with the support of corporate groups interested in some resource usage in western U.S. public lands and waters. This strongly suggests that, while the ecological data base of this movement was inadequate by contemporary demands for "environmental impact" data, there was sufficient knowledge to negate a simple "error" model of our descent into an ecological abyss. Moreover, the fact that some corporate actors supported a scientific conservation of natural resources—in the interest of long-term sustained yields for industrial utilization of these resource systems—indicates that we must modify our model of "sin" to correct for its assumption of untrammelled corporate greed as the primary basis for environmental decay. Similarly the far-ranging support for conservation-efficiency indicates there were other social groups motivated by something other than short-term income maximization in the society as well.

Second, following this early support for conservation-efficiency, this form of environmental movement declined in the 1920s and beyond. Remnants of this concern continued in some government agencies, such as the Departments of Agriculture and the Interior, but the movement organizations apparently diminished along with the decline in legislative support for conservation-efficiency. During 1920–1965 there was a continued environmental movement organized to *preserve* the pristine features of some natural habitats—the conservation-preservation movement represented by the Sierra Club and other long-standing organizations such as the National Audubon Society and the Wilderness Society (Albrecht and Mauss, 1975: 591). However, these organizations were primarily attentive to habitat features, and not to sustenance production issues as the conservation-efficiency movement had been (Schnaiberg, 1980: 10–11). That is, some social change between the 1900–20 and the 1920–65 period had occurred, which helped suppress concern about the sustenance features of the natural environment in political and social circles. Again, this contradicts a simple sin/error model, since either sin or error *once recognized*, these tales argue, would be corrected. In contrast, this historical submergence of an environmental movement and political concern suggests that either a sin/error model is inaccurate in contrast with a more complex social system perspective, or that sin and error may be cyclical features of social decision-making, and not just linear-evolutionary features that will change permanently, once addressed.

In earlier work (Schnaiberg, 1975; 1980: chs. 1, 5, 9), I have opted for the consideration of a broader and more complex social systemic perspective. This leads, in part, to a consideration of the social supports for a treadmill of production, which emerges from consideration of the dialectical relationship between economic growth and ecological structures. Regardless of which option we prefer for understanding the roots of the environmental problems, it is crucial to appreciate that *social movements have choices in how they define the processes leading to environmental degradation, and thus choices in how to create solutions to the problem.* Similarly adherents and constituents of environmental movement organizations also have choices in how they see the problem and solutions (Schnaiberg, 1973; 1980: ch. 8). This means that we should expect variations across and within movement organizations and historical time periods in dominant movement ideologies, strategies, and tactics, as well as in members' perspectives. Much of the complexity this suggests is beyond the scope of this review, which concentrates on some modal shifts in movement ideologies in the past decade and a half.[1] Obviously these trend lines conceal substantial variability within and across organizations, but they are powerful enough to warrant attention in tracing patterns of social responses to technological change and environmental impact.

One simple way of introducing this trend is to note that recent arguments about the continuity in support of environmentalism (for example, Mitchell, 1978) fail to note a change within this continuity. Apparently larger numbers of environmental adherents than in previous studies of public opinion find less incompatibility between environmental protection and economic growth. That is, "support" for environmentalism is accompanied by support for economic expansion of the treadmill of production (Council on Environmental Quality, 1980). One way of reading these data is to suggest that, increasingly, adherents of environmentalism subscribe to the "error" model of environmental problems: once we correct for the externalities of production, growth can and will continue unabated. While certainly some respected economists hold to such a view, a substantial number of dissenters in the forefront of environmental movement organizations of the past fifteen years have rejected this view at some point. Where does this new support for continued emphasis on the growth of the economy among environmentalists arise from and what does it indicate about some shifts in our environmental movements? That is, what do I turn to next?

RECENT SHIFTS OF THE ENVIRONMENTAL MOVEMENTS

One context for interpreting the roots of recent public opinion shifts noted above is in the dominant motifs of environmental movements themselves. A simple way to start is to consider the emphasis of the early environmental movements of the 1960s. These movements placed a strong emphasis on human survival—a merging of previous habitat and sustenance concerns (Schnaiberg, 1980: 10–13) of the movements' predecessors in the several conservation movements. The central claim of the early environmental movement organizations was that environmental protection would bring benefits to society and individuals; the counterpart to this message, of course, was the stress on the costs of environmental degradation. This approach enabled coalitions of environmental movement organizations to mobilize support for environmental legislation in the late 1960s.

Because of this legislation and its initial enforcement, resistance to the environmental movement grew, especially overt resistance from opposition groups that previously operated covertly (Hornback, 1974). Politically the environmental debate shifted to the costs of environmental protection; the logical corollary was, of course, the benefits of environmental degradation, though this point was less clearly stated in the early 1970s. For the environmental movement, two strategies arose in this early enforcement period: delineation of economic benefits from environmental protection, as in the pollution abatement equipment sector of employment (and profits); and cooperation in seeking more cost-effective means of environmental protection. Both of these trends led to a rather more "technical" orientation to environmental protection—shifting away from a more "political" stance. Some 1960s organizations had argued that broad human welfare was threatened, and thus production had to be regulated to protect human welfare, to ensure human survival itself.

This recent shift has been marked by a further distancing away from human welfare or social equity considerations in the use of natural resources toward considerations of productive efficiency in the use of these resources. Put another way, the emphasis of the earlier environmental movement was on natural resources as social entitlements for all citizens—with equal access to air, water, and even land. With the increased regulation of production engendered by environmental legislation, attention de facto turned more to these natural resources as factors in production—as commodities to be mobilized in production transformation, or as internalities of production rather than the previous externalities of production. Ironically political resistance by resource users (corporate and government) arose precisely because the logic of environmental regulation worked: that is, it forced an internalization of the previous negative environmental externalities of production into production decision-making. Specifically when air pollution was "forced" back into the polluting organization by the Clean Air Act, the organization struggled to externalize it once more. It often did so more fiercely than in the 1960s, when these were only hypothetical threats rather than calculable drains on profitability or competitiveness.

Following Theodore J. Lowi (1964, 1972) and Judith J. Friedman (1980), these new regulations imposed large and specific costs on particular organizations, leading to more organized patterns of resistance and bargaining around regulation. This process recently culminated in new agreements by the Environmental Protection Agency, business interests, and environmental movement organizations to an "air bubble" concept of air rights for industrial users according to which the user has effective ownership of the immediate atmosphere for emission of wastes. More clearly than most examples, this essentially confirmed the treatment of air as a commodity with only minimal social entitlements to "clean air" for individuals and nonindustrial users (for example, Tucker, 1981; Rosencranz, 1981; Alexander, 1981). Indeed, the hallmark of the middle 1970s was a bargaining stance of environmental movement organizations, which attempted to sit down with industrial opponents and government regulators and work out a *modus vivendi* that would leave all parties moderately satisfied.[2]

For the environmental movement organizations acting in the 1960s, then, the 1970s proved to be a difficult period, involving shrinkage of membership rolls as political opposition to environmental protection rose, along with the costs of such protection (Hornback, 1974). For the remaining membership and movement directorships, this appeared to be a time for bargaining, for negotiating between "ideals" of zero environmental risks and the costs imposed by such risk reduction. This led to regulatory politics in which the broad welfare argument of the 1960s was submerged beneath comparative social impact assessments or cost-benefit analyses, analyses in which distributive features were themselves often submerged (Meidinger and Schnaiberg, 1980; Fairfax, 1978). With the advent of the "energy crisis" of 1973–1974, economic hardships imposed by uncertain supplies and higher costs of energy made still further inroads into the concerns for environmental protection. Economic growth slowed, and many industrial sectors became weakened—or were seen to weaken—because of energy costs and environmental regulatory costs.

Into this partial retreat of the 1970s environmental organizations came a new social movement: the A-T movement. Spurred initially by E. F. Schumacher's work, and later by Amory Lovins's concern for "soft energy paths," and facilitated in some urban areas by a grafting onto grass-roots equity organizations loosely called the "neighborhood movement" (Pitts, 1981; Morris and Hess, 1975), this movement was the newest entry into American environmental politics. With the central social values of "peace and permanence" (Schumacher, 1973), this new movement sought to turn the problems of the energy crisis into assets for the environmental-energy movements. Given public attention to the costs of energy, one of the claims made by the A-T movement was that social sustenance could be obtained on a far lower energy budget by substituting labor for machines and could be sustained permanently through the substitution of labor-intensive renewable energy sources (solar, wind) for fossil fuels and expensive nuclear power (Lovins, 1976, 1977). For individuals and some interest groups, one of the powerful attractions of this new ideology was its apparent cost-effectiveness—living better on less money and energy. In contrast with the embattled environmental movement organizations, A-T movements could strongly emphasize the benefits of A-T, much as the early environmental movement, in the 1960s, had emphasized the benefits of clear air and water for health and recreation.

Three elements of the new environmental movement were centered on A-T. One way of distinguishing the three is to consider the multiple meanings of "appropriateness" in this movement's ideology: with respect to limited natural resources, particularly fossil fuels; with respect to the technology of production, in contrast to "inappropriate technology" or the forces of existing production; and with respect to the nature of social harmony and social equity ("peace"), or the social relations of production. To make technologies more appropriate, in the three senses above, would require somewhat different policies and somewhat different political activities (Lowi, 1964, 1972). To some extent the first required *distributive*, the second *regulatory*, and the third *redistributive* politics. In a grounded policy sense, we can think of the following respective illustrations of the three: (1) incentives to replace energy-inefficient equipment with energy-conserving equipment ("distributive"); (2) regulation of utility pricing to encourage a variety of energy conservation measures of organizations and individuals ("regulatory"); and (3) creation of windfall profits funds to provide energy cost offsets for working- and poverty-class constituents facing higher energy costs ("redistributive"). The major thrust of the movement was in the direction of the first two types of policies, while the rhetoric emphasized the redistributive goals. As a result, the new environmental movement has shifted in directions far more congruent with the shifts within the older environmental movement than most observers have noted. The nature of this shift, the factors underlying it, and its consequences are outlined next.

TECHNICAL VERSUS POLITICAL ENVIRONMENTALISM

The argument presented here is developed on two related planes. First, we will discuss how the movement has increasingly entered into or drifted into "technical" preoccupations rather than "political" ones (Friedland et al., 1977): that is, how issues of economic development rather than of welfare distribution, or of efficiency rather than of equity, have developed. Second, it will be pointed out that the recognition of this fact has been rather belated for most social science observers, who have followed the ideologies rather than the practices of the late environmental and A-T movements (cf. Mazur, 1981). To some extent, this follows the social scientific errors in determining the course of American liberalism outlined by Theodore Lowi (1979). Lowi claimed that observers analyzed the idealism of legislation processes rather than the realities of the interest-group pluralistic implementation of this legislation.

An ideal-type formulation of the transition of environmentalism in the United States can be constructed as follows. In the later 1960s, the dominant emphasis of the early environmental movement was meliorist or reformist (Schnaiberg, 1973; 1980: ch. 8). Pollution abatement to protect natural resource systems was the primary goal, and the legislation sought was primarily of the regulatory kind. What is somewhat confusing about this period is a rhetoric about fighting corporate greed and preserving the environment for "the people" or "society" that accompanied the regulatory legislation lobbying of movement organizations. Water and air pollution legislation most characterizes this regulatory policy, which imposes decentralized, disaggregated consequences and leads to some local interest group activation (Lowi, 1972: 300ff.) when the policies are implemented or implementation is anticipated. Another emphasis of the early environmental movement was on establishing a central environmental protection administration, culminating in the creation of the Environmental Protection Agency (EPA) and the Council on Environmental Quality. Both of these agencies, as well as the associated procedures for assessing federal policies under the National Environmental Policy Act (NEPA) of 1969, fit Lowi's (1972: 300ff.) category of "constituent policy," which has a more centralized, aggregated interest structure than does regulatory action such as water and air quality regulation. Thus, trade associations, along with environmental movement organizations, have been trying to "capture" the EPA for their own purposes (for example, Aidala, 1979)—either to resist regulation or to enhance regulation. Similarly the debate over NEPA and its environmental impact assessment procedures (for example, Fairfax 1978; Meidinger and Schnaiberg, 1980) revolves around the issue of whether Environmental Impact Statements (EISs) enhance or diminish environmental protection regulation—essentially asking who is capturing the processes/organizations and who is being coopted by them (for example, Schnaiberg, 1980: ch. 7).

The early politics of the environmental movement, then, involved some mixture of regulative and constituent policies, the regulative being characterized by decentralized and the constituent by centralized industrial interest groups and some political party distinctions. While neither set of policies was explicitly redistributive—and did not fit an ideal type of welfare politics (Friedland et al., 1977; cf. Lowi, 1979: chs. 8, 10)—they were somewhat less technical and more political activities in that redistributive issues were frequently raised in the context of these political battles over environmental protection. Moreover, the mass base of the environmental movement organizations dictated some public expression of concern for participatory democracy hearings on regulations and environmental agency formation (for example, Nelkin and Fallows, 1978). Thus, for example, the environmental impact assessment portion of NEPA made explicit provisions for public hearings and public review of federal policies, through the dissemination of EISs to local constituencies (cf. Friesema and Culhane, 1976; Fairfax, 1978). Initially the sense of urgency that the environmental movement felt about "survival" led to a parallel call for immediacy of state coercion in federal regulatory laws. This in turn led to industrial group formation to combat this apparently immediate threat of state sanction (fines, and so on) for the Clean Air Act and other regulatory legislation (Lowi, 1972: 300).

Two major changes occurred in the 1970s, particularly after 1974 and the "energy crisis." First, industrial interest groups found that the burden of environmental regulation could be lifted somewhat by effective organizing and litigation against federal agencies, despite the intervention of environmental movement litigators (Alford and Friedland, 1975). The costs of litigation were so high that movement organizations, with their limited resources, had to pick and choose battles carefully. The apparent result of industrial and state agency resistance to regulation was a diminished sense of immediacy of state coercion. Put most bluntly, successful resistance to the implementation of some environmental regulation led to the shift of policies away from regulation and toward "distributive policies" (Lowi, 1972), in which incentives and "logrolling" activities began to provide a larger share of inducements for production reforms. A more centralized set of "winners" and "losers" in the regulation of environmental protection gave way to the complexities of local conflicts. Workers and taxpayers often struggled with federal enforcement of environmental legislation because of the costs of compliance: loss of jobs, dramatic increases in local taxes for sewage treatment plants, and similar distributive problems dominated much of the attention to the regulatory policies of the late 1970s. In this period "jobs versus environmental protection" became a hallmark of resistance of capital and organized labor to environmental protection and anti-nuclear policies (for example, No on 15 Committee, 1976). One of the outcomes of this process was an increase in conferences between the EPA and involved communities and industries, and the emergence of pleas for moderation in environmental regulation. Environmental movement organizations less frequently ignored these pleas and more frequently entered into decentralized negotiations or bargaining over environmental regulation.

A second factor redirecting the 1970s movement grew out of a new ideological contender. The "mature" phase of the environmental movement was characterized by a mixture of regulatory and distributive policies—and most critically, by an absence of redistributive policies. But a new wave of "A-T" movement organizations promised a dramatic shift in goals and means. Schumacher (1973) emphasized that the treadmill of production was alienated not only from the realities of the biosphere, but also from the human needs of workers themselves: it served neither society nor the biosphere. The "problem of production" was thus far from being "solved," and a new production organization was sought. Hence, the promise of A-T was for greater social equity—with implications of redistributive policies and not simply of distributive or regulative adjustments within the treadmill organization (Morrison, 1980).

Yet the behavior of the A-T movement belies this promise. Indeed, despite important exceptions, the dominant thrust of A-T has been away from regulative into distributive policies, and not into redistributive policies at all. The few ventures that have been made into redistributive policies (for example, Pitts, 1981) have been highly localized, truncated, and largely unsuccessful. No organized interest groups, except perhaps political organization of the Citizen's Party under Barry Commoner's nominal leadership, have coalesced into centralized equity organizations or even stable party realignments in the legislative arena. Little effort at sustained, immediate, and powerful state coercion has been made. Instead, the priority of A-T and soft energy path movements has been the development and dissemination of energy- and resource-conserving production technologies (Schnaiberg, 1981, 1982a, 1982b; also see this volume).

Perhaps the most interesting example is in the core area of changing energy sources. In theory, the A-T movement suggests that peace and permanence can be achieved only if the treadmill of production is reconstituted into a lower level of production, with the substitution of labor for many machines and the substitution of renewable for nonrenewable fossil fuel or nuclear power. But the redistributive arguments have generally receded to the background. The movement has increasingly emphasized the sources of energy rather than the applications of energy into new forms of production (new relations as well as new technologies of production). Thus, the movement has supported federal subsidization of solar power—including the creation of a Solar Energy Research Institute as one practical compromise—as a means of moving toward renewable energy sources. This fits Lowi's traditional distributive policies arena, since no attention is paid to ensuring that the sources of subsidization are progressively financed or that the recipients of solar power are progressively allocated. What the movement apparently will accept as a successful strategy is the simple substitution of solar energy for fossil or nuclear energy.

This type of distributive policy and its accompanying politics of logrolling in Congress to ensure passage of subsidies for solar power ignores some growing evidence that energy conservation and energy substitution policies may actually increase the regressive features of American society (for example, Schnaiberg, 1981, 1982a). In effect, the A-T movement has succumbed to the requirement that it support efficiency in production, and it has in effect largely abandoned its claims for leadership of an equity campaign. Moreover, under the rubric of decentralized or "small" production, this shift to technical-efficiency rather than political-equity goals is further rationalized by the "need" to consider local interests rather than national interests. National class analyses (for example, Stretton, 1976) are largely absent from any A-T literature, despite populist rhetoric in Lovins's (1976, 1977) and Schumacher's (1973, 1979) works. This deviates from the analyses of social scientists struggling with equity dimensions of scarcity (for example, Schnaiberg, 1980; Humphrey and Buttel, 1982; Anderson, 1976).

More to the point, the A-T movement literature largely stresses equity goals but efficiency means. And it does so without incorporating a conflict dimension (Schnaiberg, 1982b) with the apparent *assumption that the anticipated efficiency of A-T means will lead to acceptance of A-T technology and therefore of the desired (but unfought-for) equity in social relations of production.* This is a transcendentalist ideology, not one leading to political organization for redistributive struggles. Indeed, although the A-T movement, like the late environmental movement, actively seeks coalitions with organized labor, there is no political program within A-T to form or activate such a political campaign to mobilize and engage in political struggle. Instead, much of the A-T effort goes into raising funds for technological innovation in energy and other resource conservation at the national or international level, or into minor political activism at the local community level around community energy policies. These often use neighborhood movements as grass-roots activists (for example, Morris and Hess, 1975; Pitts, 1981). Alternatively A-T organizations are sometimes coopted by these equity movements because of the legitimacy of the efficiency expertise attributed to A-T movements by the equity-neighborhood movements. If Lowi's (1964, 1972) arguments are correct, however, redistributive policies can only be fought for using strong interest-group activism at a centralized level of politics—as along the lines of Commoner's Citizen's Party. But the Citizen's Party platform seems much closer to traditional left-wing programs of nationalization (Stretton, 1976), with the addition of energy source changes for renewability (Commoner, 1977) rather than to a political platform for A-T production reorganization.

As further evidence of the absence of overt redistributive politics, we can point to the partial decline of interest in regulative as opposed to distributive policies by A-T movements. Incentives, not coercion, are the hallmark of A-T policy (for example, Tucker, 1981). This has been true in both energy conservation activity by the movement, as well as in energy supply policy—with the possible exception of anti-nuclear movement activity. The tie between A-T and anti-nuclear movements is extremely unclear, although it is clearer in Lovins's (1977) work than in Schumacher's (1973). While there is some compatibility between the A-T ideology and anti-nuclear movement ideology, it is much less certain that the movements have engaged in coordinated political activity to slow nuclear power. Again, one is struck with the relative absence of A-T activity in areas of regulatory conflict, let alone in leadership roles in redistributive conflicts. Despite the extensive critiques of Reagan's "reindustrialization" policies (Wolin, 1981; Dickson, 1981) which directly support the treadmill and directly oppose A-T goals, the voices of A-T leaders have been relatively silent on the resource implications. They have been virtually mute on matters of regressive surplus reallocation among social classes (Schnaiberg, 1980: ch. 5) imposed by the new tax and fiscal policies.

In short, the A-T movement appears even more depoliticized than the late environmental movement, preferring to use its resources in distributive rather than in regulative-redistributive policies. This is the hallmark of an even greater shift toward a "technical" orientation and away from the political-welfare approach that has characterized the modern environmental movements. Why does this shift appear? Some speculations follow, with a mandate for clearer research agendas on the movements.

EFFICIENCY VERSUS EQUITY PRESSURES WITHIN THE ENVIRONMENTAL MOVEMENT

Much of the analysis above is somewhat speculative, since there has been little social science research (cf. Mazur, 1981) on the overall strategies and tactics of the organized divisions of the environmental movements in the late 1970s (cf. Walsh, 1981a; Barkan, 1979; Frahm and Buttel, 1980). The interpretations above are based largely on unsystematic observations grounded in media reports and legislative histories (for example, the *Congressional Quarterly*, 1981) of environmental-energy activities. There has been much media attention (as in Talbot and Morgan, 1981) on the local or "grass-roots" efforts to build energy-conserving local policies, for example, but little systematic attention on seeking the organizational or social movement roots of such behaviors as, for instance, between origins in equity movements such as the neighborhood movement (Morris and Hess, 1975; Pitts, 1981) versus the participation and/or initiation by environmental movement organizations. Certainly A-T movement participants have pointed with pride to such grass-roots efforts, but it is not clear what role they or other movement types have played in this phenomenon. More to the point, it is not clear what successes have been achieved—and to what extent these meet the equity-efficiency duality goals of the A-T ideology.

Indeed, the movement has been credited with success in reducing energy use in the United States since 1974 (Schnaiberg, 1981, 1982a, 1982b). But this "success" was achieved by socially regressive means, leading to greater inequalities and even broader support for the treadmill of production by economic and political elites. Obviously this situation best characterizes support for Reaganomics and its "reindustrialization" goals (Dickson, 1981), but much the same position existed in the waning days of the Carter administration.

One simple illustration of this shortcoming may make the point more concretely. Both A-T and other environmental movement organizations have lauded the trend toward growth of gross national product (GNP) without growth in energy consumption—without dealing with the underlying contradictions of their position. For, while lauding this process of energy-efficiency in production (in the means of production), they have unwittingly also accepted GNP growth as a social desideratum. That is, they have accepted treadmill goals as if they were socially legitimate, thereby giving support to the very system whose means and goals they have ideologically opposed. Moreover, they have paid little attention to the de facto redistribution (Schnaiberg, 1981) that has accompanied this energy-efficiency. If anything, therefore, they have tacitly given support to the social relations of production that are more elitist, not more egalitarian. For, as in Samuel P. Hays's analysis (1969) of the early conservation-efficiency movement, it is the larger and more productive economic organizations that have been most successful in conserving energy (Schnaiberg, 1981, 1982a). Much of this conservation was a result of investing in more energy-efficient equipment, not in substituting labor for capital, as A-T movements propose. Overall, then, it would not be unfair to say that support for this kind of energy conservation is for a distributive policy, certainly not a positively redistributive one. (It is, ironically, de facto support for a negatively redistributive outcome.)

Much of this reality has escaped the attention of many social scientists, who have largely attended to the changing adherence patterns of the environmental movement (for example, Mitchell, 1978; Council on Environmental Quality, 1980), and not the systematic observation of movement organizational strategies. In ideology, then, the movements appear to be more equity-oriented than earlier environmental movements (Morrison, 1980), insofar as equity considerations enter more into the written ideologies. However, little effort has been exerted to see whether the organizational behavior follows this part of the movement ideology, rather than just the efficiency part. My own speculation is that efficiency has dominated movement activity and will continue to do so in recessionary economic times such as the present. Paradoxically, journalists and social scientists have provided many clues as to why this should be the case but have seldom followed their own clues.

Why have welfare or redistributive policies languished, and distributive (or regulatory) policies and politics grown, then, within environmentalism? Two arguments may be suggested. First, the early environmental movement arose in an historical period of great participatory movements—the 1960s (Schnaiberg, 1973). Whatever conditions influenced this participatory sweep (and among them should be noted the expansion of the treadmill that provided sufficient surplus for student enrollments and support), these conditions did not exist in the 1970s. This means that membership bases for environmental movements, including the latecomer A-T movement organizations, are smaller than in the 1960s. While nominal adherence to environmental protection values has not decreased substantially (Mitchell, 1978; Council on Environmental Quality, 1980) the constituent base has shrunk. A-T movement organizations face a different recruiting environment than did early environmentalism, including the reality that many potential constituents were already mobilized in early-late environmental movement organizations. Likewise, late environmental movement organizations and their predecessors have to fight off incursions by new A-T movement organizations. The most telling example is the old-line Sierra Club, which fragmented when Friends of the Earth (FOE) was formed as a more "political" wing, under David Brower's leadership, giving up tax exempt status to directly lobby for new legislation and engage in political support. Amory Lovins used the British FOE as a base for his own soft energy movement, indicating again the competition for constituents. One consequence of this changing membership pool, then, is that older environmental organizations struggle to maintain members, while new ones struggle to seduce members away from their predecessors, as well as to tap new pools. The movement itself, then, is engaged in internal competition that is quantitatively different from that of the 1960s at least. This has led to greater conservatism and a weaker but more encompassing ideological stance, to tap the ranks of meliorists and reformers (Schnaiberg, 1973; 1980: ch. 8).

The second factor, the changing economic and political environment, interacts with the first. Resistance to environmental regulation by industry and increasingly by government agencies facing cost pressures makes the likelihood of successful redistributive policies very low indeed. Under these conditions, movement organizations have substantial incentives to seek and support policies that will have the cooperation of many industries and state agencies. For example, the air bubble model of clean air preservation permits industrial expansion while not weakening regional air conditions (not strengthening them either, however), as William Tucker (1981) and others have described. These are essentially regulative strategies, with a small amount of distributive policies attached (as in de facto or de jure support for pollution abatement equipment manufacturers).

Where the first and second set of constraints intersect, they produce a politics of commodities (Schnaiberg, 1982a, b). While the ideology of A-T (and some environmental predecessors) is one of resource entitlements and that of early-late environmental movements one of resource preservation, in the late 1970s conditions shifted in the United States to produce attentiveness to the commodity dimensions of resources. Thus, efficiency concerns dominate, since they are directly linked to commodity production (and presumably to commodity distribution, though more weakly). In this one area membership and elite coordination is possible, since it is presumably in "everyone's" interest to maximize resource-usage efficiency in production. In this arena, therefore, technical politics have predominated, and this has shunted welfare politics aside. Interestingly, following Hays's (1969) analysis, it is also the arena in which conservation movements have traditionally operated. The redistributive element has been largely absent from most of the history of environmental movements, despite vaguely populist rhetoric (Schnaiberg, 1980: ch. 8; 1981; 1982b).

CONCLUSIONS: THE FUTURE STRATEGY OF ENVIRONMENTALISM

Contrary to Hugh Stretton's (1976) expectations, it is not likely that the near future will bring a sudden upwelling of A-T or any other environmental movement expansion of redistributive politics. Following James P. Pitts's (1981) analyses, it is more likely that a coalition around equity issues will start building when equity movements see a stake in it and reach out to efficiency-environmental movements. The reverse situation seems unlikely because A-T and other environmental movements are simply not welfare-oriented to the degree that a stable, sustained coalition-building effort is possible. A-T groups will be more likely to take a staff, not a line, role in the equity-efficiency politics that Stretton (1976) envisages. They will form the advisory groups and perhaps share in the lobbying efforts and some of the technical-organizational tasks of fund-raising and the like. But they do not appear eager for the raw political struggles historically associated with welfare politics.

This forecast poses a challenge for social scientists. We have devoted far too much attention to the vagueness of attitudes attached to the movement, and far too little to the recruitment and socialization and performance of movement constituents. While this approach was understandable in the 1960s when the participatory base was so wide, this justification is not valid today when the environmental movement must use its constituents much more sparingly and efficiently. Only a vague impression of the directions of environmentalism can be provided since the data are limited. Interestingly in these tight times for research, it may be a blessing to note that intensive, participant (or non-participant) observation of movement organizations may be much more feasible through labor-intensive methods of research than public opinion surveys. Maybe this is our own version of "appropriate" social science technology.

NOTES

1. Some earlier writings touch on the divisions within the environmental movement (Schnaiberg, 1973; 1980: ch. 8; 1981), although these are all based on very limited data on movement organizations. Most "environmental movement" data in the modern period are on movement adherents, using mass public opinion survey data (for example, Hornback, 1974): we lack longitudinal data on movement constituents, the actual participants in movement activities (cf. Mazur, 1981).

2. Much of this activity seems to reflect the processes and outcomes noted by Alford and Friedland (1975: 455–64) in their treatment of "bureaucratic participation without power." Parallel analyses by Lowi (1979) suggest much slippage between legislative principles and administrative practices when such negotiations occur.

III
The Legal Response

This section on legal responses contains two key papers. Professor Thompson's paper is a socio-legal analysis of structural conditions affecting the application of laws concerning environmental harm. It focuses primarily on the West Coast fisheries, and it concludes that conditions in the industry and in the relevant civil service sector structure interpretation and application of the law into what Thompson considers to be a more appropriate negotiation/bargaining mode.

Professor Elder strongly takes issue with Thompson, maintaining that negotiation is somehow "sleazy" and suggesting that the public interest is not best served by such a mode, understandable as it may be. Given the unequal power and accountability of the civil servant regulator and the industrial spokesperson, he believes it is imperative that the public interest be represented through public forum and debate before policy decisions are made in all such cases.

Professor DiSanto amplifies Thompson's thesis by discussing several kinds of real-world problems that limit technical or scientific decision-making, including political factors, value judgments, jargon, and specialized funding agencies. He concludes with the observation that while our concern with equity is most often operationalized in terms of attempts to equalize "condition" (or rewards), we might be better off to concentrate our efforts on increasing the opportunity for mobility among the growing numbers of positions while leaving their reward structures unequal.

Professor Carson's paper deals authoritatively with health and safety legislation in the context of offshore oil development, primarily in the North Sea. It persuasively argues that such legislation, or the lack of it, is the product of an extremely unequal version of the bargaining mode that Thompson discusses. Profits come before safety, and hazard is the price of profitable and early development. In this particular instance, Carson argues that the British public interest has been overwhelmed by private and public vested interests with economic/political incentives and motivations. Government and industry have worked hand in hand to create and support safety standards that are both lax and demonstrably and tragically inadequate.

In his comment on Carson's paper, Professor Pratt expands on the Carson thesis, introducing considerations that are more particularly Canadian in nature, while echoing Carson's warning that tragedies like the *Ocean Ranger* will continue to occur as long as technological gambles are made with the premature support/acquiescence of a government that is more concerned with its own potential political and fiscal gain than with workers' lives and safety. The role of Petro-Canada in this context becomes doubly problematic as an industrial giant and simultaneously a creature of federal government oil development policy in Canada.

3

Bargaining with the Environment: The Limits of Legal Regulation

ANDREW THOMPSON

If the breadth of concern about science, technology, and the legal process being expressed in academia is a sound indicator, there are good reasons for examining traditional legal responses to environmental regulation. In philosophy, A. K. Bjerring and C. A. Hooker (1980) question the adequacy of empiricist-formalism as a foundation for science, advocating instead an "actional-process" theory, with emphasis on a normative vision of science in which "the broader focus for all policy becomes the selection of processes for change within the constraint of fundamental uncertainty."

Under Professor Hooker's stimulus, a national conference at the University of Waterloo has provided the Social Sciences and Humanities Research Council with an agenda for a concerted research effort to deal with issues of science and technology from a vantage point of human values and social policy (Hooker, 1980). The emphasis here is on policy formulation and decision-making, based on exploration of fundamental questions such as "the moral and social limits (if any) to scientific investigation and technological development" and "concern to relate support of science and technology to the support of individual and group development."

This concern about science and technology is not exclusively the preoccupation of philosophers and social scientists. Parallel concerns are being expressed about legal positivism in response to scientific and technical problems. In 1978 the Science Council of Canada established a committee on Science and the Legal Process comprised of an equal number of scientists and lawyers to study and report on the interrelationship between scientific investigation and social policy and how the legal process deals with scientific uncertainty and controversy. The work of this committee, under the chairmanship of Dr. David Bates, a University of British Columbia (UBC) epidemiologist, is now in final draft stage under the title "Certainty Unmasked: Science, Values and Regulatory Decision Making." This report is based on component studies of the legal problems associated with new biological developments (prenatal diagnosis, genetic screening, recombinant DNA techniques, biotechnology, advances in reproductive technologies), of how government departments utilize science and respond to scientific and technological issues (Campbell and Scott, 1978), of the role of public inquiries in the assessment of scientific and technical issues (Salter and Slaco, 1981), and of modes of dispute resolution used by the parties to scientific and technological disputes (adversary proceedings, mediation proceedings, consultative modes).

The legal profession has its own preoccupation with this subject. The Law Reform Commission of Canada has a research project underway to examine the adequacy of sanctions and compliance policy in the field of administrative law, with a heavy emphasis on government regulation in fields where scientific and technical questions predominate (Eddy, 1981). The Canadian Institute of Natural Resources Law at the University of Calgary recently held a workshop on "Environmental Law in the 1980's—A New Beginning," the stated purpose of which was to reexamine fundamental approaches to the regulation of technological processes that affect the physical environment. "In a word, the conceptual bricks and mortar of environmental law may not be suited to current demands on it," said the organizers of the workshop.

Last, but possibly not least, are studies undertaken by the Economic Council of Canada under its mandate from Canada's First Ministers to explore the prospects of "deregulation." Westwater Research Centre at the University of British Columbia was commissioned by the Council to carry out the environmental regulation component of the Reference. For this purpose Westwater commissioned six case studies and three theme papers.

In the case studies, the attempt was made to understand how much environmental regulation is costing private industry and government. How are these costs distributed between private and public sectors? Can the costs and benefits of this regulation be assessed? Do overlapping and conflicting government jurisdictions cause serious inefficiencies? How effectively are knowledge and information being used by regulators? How well is the public involved in regulatory processes? Lastly, what can be done to improve the regulatory processes?

The theme papers were intended to deal with environmental problems of a pervasive nature. Professor Donald N. Dewees, an economist at the University of Toronto, sought to explain why public land managers have been so reluctant to adopt the solutions proposed by economists whereby pollution rights or charges would be used instead of prohibitions and penalties to induce more responsive behavior on the part of industry. John Swaigen, a lawyer with the Canadian Environmental Law Association in Toronto, examined the many legal and political difficulties that beset our compensation laws so that those who suffer environmental damage are often without remedy. Westwater Research studied how major projects—in particular, the British Columbia Hydro Revelstoke Dam project—are assessed and reviewed, and how their impact is managed and monitored.

Environmental Regulation in Canada (Thompson, 1981) presents the following conclusions:

The administrative cost of environment regulation is minimal, in both the private and public sectors;

Operational costs are extremely difficult to quantify but seem to be within the limits reported for other countries and not excessive in general;

Much of these operational costs is redistributed from the private sector to the public sector through the operation of taxes and allowances;

It is likely that in general the benefits of regulation substantially exceed costs, but this cannot be proved;

At the margin it is usually not possible to establish that the benefits of a given control measure will outweigh its costs;

Overlapping and conflicting jurisdictions do not impose serious inefficiencies and sometimes, as in the case of fish and forests, are beneficial;

The regulatory system fails to use knowledge and information effectively;

The public is inadequately involved in the regulatory process;

Legal enforcement procedures are ineffective;

Bargaining characterizes the regulatory process.

The pervading tone of these conclusions suggests that we are caught in a process where lack of information, misinformation, and uncertainty have become characteristic and normal. It seems to be in the nature of man-caused environmental effects that dealing with them is knowledge-dependent in a special way. By definition, they are the unintended spillovers from purposeful activities, usually commercial ones. As such, they must be identified and traced—their potential for harm must be measured and in many cases they cannot be clearly connected to any particular activity. Physical effects on land and water are difficult enough to monitor. But biological effects are more complex, and most are not understood at all in terms of a total ecology. Yet it is injury to natural life, including human beings, that is most often the focus of environmental concern.

Ideally cause-and-effect relationships between the questioned activity and the apprehended environmental harm must be established. With that knowledge the questions of alternatives and mitigation can be addressed. For this purpose a thorough understanding of available technologies is required. But since we are talking about government regulation of the questioned activity, the question of social utility must be answered as well. Given our understanding of cause and effect and of available alternatives and mitigative technologies, should the activity be regulated at all? Some attempt must be made to calculate the tradeoffs between the benefits of the activity and the costs of its regulation.

Nor does the need for inquiry stop here. Obviously an analysis of costs and benefits in a broad social sense tells little about the impact on individuals. The distribution of costs and benefits must also be foreseen, because equity and fairness are major societal values for which governments carry special responsibility. Finally, the alternate methods of accomplishing regulation, and the costs of these alternatives, must enter into this ideal decision-making.

These uncertainties are not unique to environmental regulation but instead seem to be typical of the limitations faced whenever the need to regulate is confronted by applications of science and technology in a human context. From the point of view of decision-making and implementation, government regulation in this field follows traditional positivist forms. Standards are legislated, and prohibitions are enacted to be enforced by procedures of a penal nature. But in practice, these forms are displaced by negotiating and bargaining. This is the conclusion of the case studies. Not only is the mode one of negotiation and bargaining, but this mode is "likely the only system that can work in present circumstances" and "the only one that can cope with the uncertainties of environmental regulation in a practical way" (Thompson, 1981: 50–51).

Such a conclusion runs counter to the intuitive view that bargaining is something which goes on because regulators are too weak or too venal to enforce environmental standards. Since most environmentalists argue for stricter standards, more rigorous enforcement, and stiffer penalties, this finding requires further analysis and justification. The purpose of this paper is to examine the limitations of science and technology in the context of government regulation, the reasons the traditional forms of regulation by prohibition and penalty are ineffective, and the limitations and opportunities of a bargaining approach to environmental regulation.

THE LIMITATIONS OF SCIENCE AND TECHNOLOGY IN AREAS SUBJECT TO GOVERNMENT REGULATION

Brought to a head by the recombinant DNA controversy, questions about the legitimate role of science are preoccupying philosophers of science today. In the DNA case, the question was whether even pure scientific research should be regulated in the public interest. More often the questions concern the application of new bioscience technologies in fields such as prenatal diagnosis or genetic screening. In general, studies of science policy show that, on the one hand, the expectation of science as a source of solutions to human problems is unrealistically high and, on the other hand, the use of science as a basis for decisions is unexpectedly low. It is remarkable that three recent studies, two Canadian and the other American, reveal, contrary to expectations, that scientific research and analysis play relatively peripheral roles in government decision-making in such fields as health and safety (Crandall and Lave, 1981). It was not that regulators were not confronted with scientific controversy. In the U.S. study, institutional restraints were identified as limiting to science. In particular, legal requirements stood in the way of scientific decision analysis. Uncertainty also undermines the use of science. "Since scientific evidence remains uncertain, regulators need a way to deal with this uncertainty just as much as they need better scientific evidence" (Crandall and Lave, 1981: 15). Without systematic methods for taking account of uncertainty, the regulator tends to obscure the fact that scientific controversy exists.

This tendency in turn reinforces the public's unrealistic expectations that science and technology can supply the answers. Closely related to the uncertainty problem is the fact that scientific conclusions, when they are applied to solving human problems, invariably incorporate a range of value judgments. If these are acknowledged, the tendency of the pragmatic professional is to say that, since the issue involves value questions, it might as well be confronted as a political choice without the need for an expensive and time-consuming scientific analysis. Alternatively the regulator may ignore the value question and carry forward the pretense that his or her decision is a purely technical one. In this case it is better not to pursue the scientific inquiry too far!

It is beyond the limits of this paper to pursue very far what can be learned about government regulation from an analytical style that explicitly recognizes the limits of science. The following framework is drawn from the recommendations in the draft report of the Science Council Committee on Science and the Legal Process.

Sources of Concern

The 1960s brought an awakening of social consciousness largely because the principles and mode of thought of the science of ecology were transferred to the understanding of the person on the street (for example, Charles A. Reich, *The Greening of America*, 1970). This awakening entails an expanding sense of responsibility for the consequences of actions.

There has been an explosion of knowledge about cause and effect and about interdependencies within and between systems, reinforcing and feeding the expanding sense of responsibility (for example, epidemiology, tracing of chemical pathways).

New orders of scientific and technical problems are arising from the explosion in technological application of new science. Examples might include: human genetics, the so-called genetic engineering; biotechnology; multiplication of new drugs, chemical additives, pesticides, and so on; nuclear science; "chip" technology—microprocessors, and the like; and universal communication systems.

There are new problems of scale as modern technological applications seem to require ever larger projects commanding more and more natural and human resources and visiting wider and wider impacts on the human and natural environments.

Impact on Legal Processes

The expanding sense of responsibility for actions has: increased the demand for regulatory intervention by governments; reduced emphasis on private property interests as a basis of accountability for broad "public interest" impacts; extended the range of duties of care and of foreseeability in tort liability; increased the emphasis on abatement and compensation rather than fault and penalty; and liberalized the rules of standing.

The new scientific knowledge concerning cause and effect and interdependencies results in: questioning the adequacy of information systems used by industry and government; questioning the onus of proof (for example, the onus of establishing public interest should rest on the party introducing the new technology); and questioning dispute resolution procedures (for example, use of experts in adversary proceedings).

The new orders of scientific and technical issues result in: challenging established value assumptions about human worth, public goods, and so on. This in turn results in a demand for recognition that technological issues implicitly involve value questions; demand for openness and accountability in regulatory processes so that value assumptions are not obscured by technological findings; demand for lay participation in professional and other expert deliberations; demand for increasingly formal supervision of regulatory procedures (appeals, judicial review, and the like); and a demand for courts to be more aware of the true nature of scientific determinations about technology.

The new scale of technology results in: rejection of reductionist empiricism (for example, formal cost-benefit analysis as the sole or principal test of public interest); distrust of narrow specialization in government mandates and regulatory agencies; insistence on a holistic approach to issues (for example, justification of a project requires consideration of a wide range of alternatives); demand for broader departmental mandates (for example, B.C. Forest Act); and demand for large-scale public inquiries with broad terms of reference.

Making Legal Processes More Responsive

A case is made, not for more government regulation, but for better regulation. Regulators need to be better informed both in the sense of understanding the nature of their responsibilities and the contributions science can properly make to discharging these responsibilities; and in the sense of ensuring that the scientific and technical data and knowledge of functional relationships concerning the issues for which they are responsible are available and properly used when they discharge their responsibilities.

Institutional structures (regulatory agencies and the like) should be given mandates that require them to recognize broad cause-and-effect relationships and interdependencies. Regulatory decision-making should be more open to ensure both accountability and broadly informed decisions as to value questions. Rule-making procedures should be more open for the same reasons. In addition, there should be greater scope for judicial review of government decision-making and for a judiciary more aware of the nature of scientific and technical controversy.

Compliance mechanisms should recognize the fundamental uncertainty in applications of science and technology and should be part of an integrated approach to decision-making that concentrates on achieving goals of social policy that are themselves evolving. Above all, social policy must be explicitly formulated and reformulated so as to provide the goals for which decision-makers strive in science and technological controversy.

In summary, there are limitations on the application of science and technology in government regulation. These are behavioral in some cases, as when expectations are unrealistically high, and institutional in other cases, as when legal requirements constrain analysis into artificial patterns. In other cases the difficulties are weaknesses of the government regulator, as when uncertainties are inadequately confronted, value judgments are obscured, or information systems are defectively used.

The Ineffectiveness of Prohibition and Penalty

There is long-standing debate about how appropriate prohibitions and penalities are as instruments for gaining compliance with regulatory goals of government when the offending conduct lacks the moral turpitude normally associated with criminal conduct. On one hand, it is argued that the criminal law and procedure are placed in disrespect when employed indiscriminately as compliance mechanisms without regard to fault or misdeed (Eddy, 1981). They are seen to be too heavy handed, to encourage unnecessary evasion stratagems, and to operate haphazardly because of the safeguards (loopholes) that are designed for the serious offenses with which criminal law is intended to deal. Or, if prosecution and penalty are successfully imposed, the inappropriate criminal nature of the process is seen as vindictive, stimulating behavior resentful of law and order. On the other hand, regulatory requirements are often onerous, penalties must be proportionate, and procedural safeguards are necessary to protect the citizen from an overreaching bureaucracy. Besides, the criminal law and procedure are there and ready to be used. So why not use them, especially since a substantial number of regulatory offenses do cross the border into truly criminal territory.

The Law Reform Commission of Canada, in its *Report 3, Our Criminal Law,* takes the position that "regulatory offenses should be excluded from the Criminal Code, should involve no stigma, and should not be punished by imprisonment" (1977: 18–19). An exception would apply in the case of intentional and serious breach of a regulation amounting to a real crime, such as fraudulent violation of weights and measures regulation.

But these arguments about the appropriate use of criminal law are not the arguments one hears with respect to environmental regulation, particularly in relation to pollution control. Rather, when the emotional rhetoric about the evils of pollution dies down, the ensuing debate will be characterized by frustrated consensus that regulatory controls are not working. The environmentalists blame weak and uncommitted enforcement agencies, while the regulators plead technological and economic constraints. The economists believe that they have the solution if only discharge or pollution rights could be traded in a marketplace, knowing all the while that two decades of theory have failed to achieve a viable pollution rights system in Canada or the United States.

Clearly this frustration pervaded the six case studies on environmental regulation. If enforcement was not working, what elements did characterize the regulatory process and why?

> [T]he reality is that the rules of environmental regulation are never clearly stated or certain, except in a purely symbolic sense. Instead the norms of conduct are the subject of negotiation and renegotiation between the regulator and the regulated right down to the moment of compliance or noncompliance.

> In this sense, rules stated in statutes or regulations are merely points of departure for negotiating modifications of behavior; and "compliance" or "noncompliance" means "agreement" or "disagreement." Only if there is an ultimate disagreement is the enforcement procedure utilized, and even then its role may be but another step in a drawn out negotiation process, of little more significance than a variety of other weights such as tax concessions, that the regulator may bring to bear. (Thompson, 1981: 33)

The overview analysis suggests that this resort to bargaining is necessary because it is the only approach that can deal rationally with the pervasive uncertainties besetting environmental issues. Criminal law enforcement procedures appear to be unsuited to environmental regulation except in the case of intentional discharge of known dangerous substances (midnight dumping). Perhaps this requires further explanation.

In all branches of obligation the common law requires a sufficient degree of certainty in specification of defaulting conduct so that a finding of fault can be based on known criteria. This is as necessary for the adjudicating tribunal as it is fair to the alleged offender because otherwise the tribunal can rely only on unbridled discretion without any basis for an objectively determined judgment, which is the essence of impartiality in such decision-making. Where criminal sanctions are involved, the required standard of certainty is at its highest for obvious reasons. This means, in the case of pollution, that the ingredients of a pollution offense must be specified with certainty in advance of any prosecution.

But this is not all. It is of the essence of the criminal law that any offense be of general application to a class of potential offenders. It is a truism that punishment must fit the crime, but we do not tailor the crime to fit the individual. Consequently the offense that is specified with certainty must be one that applies to a class of polluters and not one that is separately defined for each individual polluter. Finally the criminal law, as the heaviest hand of the state, depends more than other branches of the law on a rational relationship between apprehended harm and the weight of legal consequences.

Now we can define the series of dilemmas confronting the effort to use the criminal law and procedure to sanction pollution offenses. First, because of our limited knowledge of natural systems and of the effects of introducing new substances or initiating new processes into the environment, it is virtually impossible to define the ingredients of offenses in a way that is specific to the harm done. It is generally acknowledged that pollution standards ought to be ambient standards that are indicia for actual harm done to the specific environment affected. But because our science and technology usually cannot provide these ambient standards, we can achieve certainty in creating offenses only by specifying point discharge standards—that is, so many parts per million of the pollutant measured at the end of the stack or discharge pipe. If such point discharge standards are specified for a particular operating plant such as a pulp mill, we offend the principle that crimes must not be tailored to individuals. If we then fall back on point discharge standards of general application, we produce either meaningless or irrational results. The meaninglessness is usually disguised by a statement that the standards are to take effect only as guidelines. Then the bargaining begins to achieve either an exemption from the standards for a particular plant, as was done at the Kitsault Mine discharges into Alice Arm in British Columbia, on a specification of new standards to be tailored to the specific plant operation by the terms and conditions of a pollution control permit or license. Those charged with this task—making rules for a particular plant operation—are bound to see their task, not as creating and enforcing criminal law, but as ongoing regulation of an industry. Because the pollution is part of a beneficial production technology and because control of pollution is itself an evolving technology, rationality requires the regulator to make tradeoffs both between the costs and benefits of the abatement procedures and between what can reasonably be accomplished now and what can be left for a future date.

The chief example in Canada of a pollution offense that does meet the tests of certainty in statement and generality in application is §33(2) of the Fisheries Act which makes it an offense to be responsible for introducing a deleterious substance into waters frequented by fish. This is the paradigm case of the pollution problem. The general nature of the prohibition and its sweeping definition of unlawful conduct make it virtually impossible to carry on many normal occupations without violating the law. Both logging and placer mining are examples. Operating a pleasure craft with a flushing head is another. In practice, the conduct that constitutes an offense is determined by a fisheries enforcement officer in each particular case. All criminal law prosecutions involve some discretionary intervention by police officers and prosecutors, but none permit the enforcement officers to define the offense from case to case in the way that occurs under the Fisheries Act. In consequence, this major environmental offense that meets criminal law standards of certainty and generality on the face of the statute fails to do so in practice. The resulting discrimination and unfairness place the Fisheries Act and fisheries enforcement under broad condemnation by those in the forestry and mining industries.

To sum up, the knowledge gaps about environmental cause and effect, the uncertainties about abatement technologies, the difficulties in calculating costs and benefits, and the persistence of the environment in being unique at any given time and place combine to produce a state of affairs where it is virtually impossible to specify a pollution prohibition and penalty that will meet criminal law tests of certainty, generality, and rationality. Hence a bargaining mode!

This matter of enforcement cannot be left without making further distinctions. The standard case is considered to involve prevention of harmful pollution by substances produced as part of a production process of goods that are in the public interest to produce. This kind of pollution can be conveniently labeled "process pollution." It is in dealing with this process pollution that the bargaining approach seems the preferred method of gaining optimum control measures. A study is presently being performed for the Law Reform Commission of Canada to evaluate whether control measures could be negotiated more efficiently and with better results if licensing, prohibitions, and penalities were replaced by contracts between the government and the polluter specifying compliance requirements and enforcement measures.

Other types of pollution offenses might also be defined for enforcement purposes. In addition to process pollution, "casualty pollution" and "known toxics" can be identified. Casualty pollution is different from process pollution in that a pollution control measure has been specified and is in force (for example, by contract), but some failure in containment or neutralization of the pollutants occurs. Failure may be due to unforeseen causes or to negligent or even willful causes. If unforseeable, contractual sanctions would be adequate. If failure is due to negligence, then in addition to contractual sanctions it is appropriate to treat the negligent or willful polluter in the same way as any other agent who negligently or willfully fails to carry out regulatory duties. Even so, as the Law Reform Commission of Canada advocates, such a regulatory offense would not be treated as a breach of criminal law, no criminal stigma would attach to the offender, and the legal defense of "due diligence" would be available.

"Known toxics" signifies a pollution offense that is truly criminal in character. When substances known to be dangerous to human health and safety are intentionally discharged into the environment, the offense has equivalent elements to many other Criminal Code offenses where the public is endangered by conduct that is reckless or intentionally harmful.

This decriminalization of environmental law cannot easily be justified by an anthropocentric argument that harm to the environment is morally wrong only when there is danger to human beings. But it may be justified as the more effective way to achieve an environment that will be safer for all life on the planet.

THE BARGAINING PROCESS

If bargaining characterizes the environmental regulation process, it is as necessary that realistic analysis be focused on this bargaining mode as that it be focused on the limits of science and technology and on enforcement methods. A host of questions have not received adequate study in Canada, and so the scope for research is almost unlimited.

Murray Rankin and Peter Finkle (1981) warn of a danger in recognizing environmental regulation as a bargaining process. They argue that it is possible to move too quickly from a descriptive to a prescriptive analysis, with the consequence that reform consists merely of improving the bargaining process. "That would be to say uncritically that what is ought to be." Their central thesis is that bargaining is appropriate up to the point of setting environmental standards, but that once standards are set they must be enforced with the same diligence as the state enforces the criminal law. But if the setting of standards which can meet tests of certainty, generality, and rationality is the goal that is so elusive in the field of environmental regulation as to require a bargaining mode, it seems logical that concentrating analysis on the bargaining process is not only the essential first step but also the most fruitful step that can be taken toward improved regulation.

Bargaining itself is a subject of long-standing relevance in fields like international and industrial relations and marketing, and even in contract law. It is a topic in a more abstract sense in economic and political science theory and in systems approaches to analysis. What is new in the present case is the characterization of the bureaucrat in Canada as bargainer.

The ordinary typology of bargaining assumes tradeoffs between private sector parties, each motivated by profit and loss considerations in the conventional marketplace sense. It may be arguable whether the bureaucrat as bargainer differs from this typology in kind or merely in degree, for private sector bargainers are often representatives of interested parties rather than the parties themselves, and the interests represented in the private sector are often diffuse and difficult to measure in ordinary marketplace terms. In any event, the differences are substantial and warrant careful analysis.

The Bureaucrat as Representative

The civil service tradition under British parliamentary government portrays the bureaucrat as a professional faithful to his or her parliamentary mandate no matter what policy is being pursued or what political party is in power. As a representative of a parliamentary mandate in a bargaining process, the bureaucrat should have clear and explicit bounds within which to exercise discretion. Instead, we have seen that the very reason why the bargaining mode is prevalent in environmental regulation is that pervasive uncertainty obscures policy goals and the means of attaining them. Bureaucratic discretion in bargaining is exercised not pursuant to parliamentary mandate but in spite of it.

Nevertheless, the image of the robot bureaucrat implementing government policy but never making it persists. For example, the federal Department of Environment's guidelines for bureaucrats who appear as witnesses before inquiries insist that they speak only facts and personal opinions and that in no wise do they speak to questions of departmental policy.

The extent of discretion exercised by regulators in pollution control is enormous. Otherwise, one cannot account for the fact that studies invariably reveal a record of non-compliance with emission standards and pollution control permits. Agents for private parties also exercise discretion when negotiating, but their mandates are usually clearly bounded, being framed by a body of well-established legal principles whereby a principal can control his or her agent and even escape third-party liability if the agent exceeds certain limits of authority. The very rare cases in which a regulator is restrained by a court in some aspect of bargaining with industry testifies to the fact that similar clear bounds and judicial restraints do not apply.

The essential difference between public sector and private sector representatives seems to lie in the area of mandates. There are good reasons for the difference. Just imagine the two sides to an issue concerning the installation of a costly pollution control apparatus in a fish plant. On the private side there is intimate familiarity with the industrial processes and therefore a clear understanding of the technological problems involved in the installation. Costs can be confidently calculated. But, most importantly, the mandate bounding all judgment is normally the benefit-cost calculations of the company. On the public side, none of the technological or cost data are readily available or clearly reliable, and the mandate is as likely as not confusing and even contradictory in its elements. That mandate derives not just from statutes and regulations, but from political pressures and pragmatic considerations as well, nowhere as neatly defined as they are for the private sector in the balance sheets of industry. Perhaps this difference in mandate is more a matter of degree than of kind, because industry, too, faces medium and long-term objectives that are not entirely reflected in profit and loss statements.

But clearly the bureaucrat as bargainer has a much more difficult task than his or her counterpart in the private sector owing to the obscurity and complexity of the bureaucratic mandate. This difficulty is not lessened by the pretext that the bureaucrat is a faceless servant of the will of Parliament.

There is another context within which the bureaucrat functions as representative of the public interest. In those cases, particularly in western and northern Canada, and in the offshore regions, where governments enjoy extensive ownership of natural resources, the mandate of the bureaucrat is much closer to that of the bargainer in the private sector. The professional forester in the Forest Service of British Columbia has an outlook and responsibilty similar to those of his or her counterpart in MacMillan Bloedel as each bargains to maximize the value of forest products over time. Environmental protection measures in the form of fire suppression or insect control can be directly motivated by the simple fact of public ownership. Perhaps the interests of environmental protection might be better served in Canada if more emphasis were placed on the role of government as owner of resources than as regulator of resource users. Water pollution control can be viewed as a self-interest measure of the owner of the water resource (in British Columbia the beds of lakes, rivers, and streams, and the fresh water fisheries are owned by the provincial government) instead of as a regulatory responsibility. This difference in approach could lead to clearer policies and more forceful bargaining on behalf of the public sector.

The Bureaucrat's Bargaining Tools

In the field of pollution control, the chips that the bureaucrat can lay on the table are entirely different from those in the hands of the private sector bargainer. The bureaucrat's chips fall into three main categories: financial incentives or other kinds of support; permits or licenses; and prohibitions and penalties (Doern, 1981). The first category—incentives—seems to involve less discretion than the others because incentives are usually spelled out in explicit terms in granting or taxation statutes and regulations. The bargaining in this case usually occurs at the political level when industry makes its case for special treatment as in Ontario where incentives have been created for modernizing pulp and paper mills.

The granting or withholding of permits and licenses, and the opportunities they provide for imposing and enforcing terms and conditions, are the main bargaining tools of provincial pollution control agencies. On the federal side, bargainers are armed mainly with the ability to initiate prosecutions under the Fisheries Act. Under §33.1 of the act there is provision for the equivalent of a permit system with respect to works that result in or are likely to result in pollution or in the alteration, disruption, or destruction of fish habitat, but the Department of Fisheries and Oceans has thus far used this provision sparingly.

The effects of the provincial permit system and the federal prosecution system have not been that dissimilar. In each case negotiations take place concerning the upgrading of existing plants and facilities, so as to reduce pollution, and concerning the design of new plant and facilities, so as to avoid pollution. The two systems merge to the third category, when bargains are broken because the provincial regulator usually relies on the threat of prosecution to secure compliance. He or she can revise permit terms and conditions and threaten the cancellation of the permit, but this draconian measure seems to be unused. Even under the Fisheries Act the federal regulator may revise terms and conditions and reverse understandings about immunity from prosecution. In both cases, bargaining may, in fact, take place right through to the initiation of prosecution and, to a degree, thereafter.

Both the permit and prosecution modes seem to be conditioned by the aura that pollution is a crime and that problems should be dealt with authoritatively and punitively. Professor Franson and I are collaborating in a study of negotiated contracts as a means of implementing pollution controls. The idea is that flexibility would be gained not only in the design and management of pollution control measures but also in the means of ensuring compliance.

How the Marketplace Functions

There are few detailed studies of the institutional nature of the bargaining system in pollution control. For some reason those that do exist are concentrated in British Columbia where the Pollution Control Branch and the federal Department of Fisheries and Oceans have been the subjects of behavioral and institutional research (Dorcey, 1981; Sproule-Jones, 1981). Generally these studies show systems that involve hundreds of day-to-day dealings between government and industry that bespeak a well-understood set of roles and expectations, with an occasional flareup when some incident such as a fish kill or a media campaign disturbs the status quo. Despite the calm of accommodation, there are persistent complaints, both by industry and environmental groups, about jurisdictional overlaps, inefficiencies, and failures to meet compliance expectations.

But, while this research in British Columbia has identified the bargaining nature of the regulatory process and described it in operation, the bargaining itself is analyzed in only one of the studies where "public choice" theory is applied (Thompson, 1981). Much remains to be done.

Bargaining and Due Process

Finally any analysis of the bargaining process in environmental regulation must consider how due process is to be observed. Canadian courts are presently enunciating a "duty of fairness" in administrative decision-making that is broader in application than the well-known "principles of natural justice" (see Island Protection Society v. the Queen [1979] 4 W.W.R.1). The articulation of the constitutional

Charter of Rights may lead to an increasingly interventionist stance on the part of the judiciary as they perceive an increasing responsiblity to stand between the citizen and the state. If the bargaining nature of environmental regulation is more explicitly acknowledged, the result may be a heightened awareness that procedural fairness is a legal requirement.

On the side of the public interest in a pollution-free environment, there will also be due process requirements to ensure that the regulators remain responsive to their mandates in the bargaining process. When it is clear that there are limits to scientific and technical solutions and that there are basic value questions at issue when tradeoffs between costs and benefits are made, there is an obvious requirement for forceful representation of the various public interests involved in the bargaining.

The final note on which to end this paper makes the case for public participation in the environmental regulation process in the following way:

> [S]ince environmental regulation is of necessity judgmental, and negotiations and bargaining characterize the process, public participation is seen to provide a necessary aid to decision-making. It can inform decision makers about impacts and values, it can deter the negotiators from resorting to bluff and threat, and it can instill public confidence that balanced decisions are being sought. (Thompson, 1981: 45)

4

Counterpoint: The Politics of Bargaining with the Environment

PHILIP ELDER

There is something sleazy about the notion of governments bargaining away environmental quality with powerful economic interests. Backroom deals; favors for political friends; Minamata disease; the Sudbury moonscape; and acid rain. Yet rhetoric aside, is there any choice? Do not powerful vested interests compel bargains to be struck? Can any one interest group (whether environmentalist or industrialist) claim that its interest is paramount? And, as Andrew Thompson's excellent paper argues, with tremendously complex ecological interdependencies and scientific uncertainties, how could we ever achieve the certainty, generality, and rationality of standards for the enforcement of quasi-criminal law? At one level, of course, it is obvious that government decision-making is political, and hence involves compromise and bargaining. Nor should it be otherwise. But Professor Thompson is saying something more. The political bargaining process in some policy areas can result in certain, general, and rational criminal prohibitions, but he argues that these are not suitable for the plenitude of different physical circumstances in which each plant or mill finds itself.

Given that there is no perfect answer, why cannot the environmental bargaining process produce license standards unique to each mill which, if exceeded, almost automatically result in quasi-criminal proceedings with financial penalties upon conviction? If this results in greater public notoriety than speeding convictions and fines, it may be because society holds the deliberate exceeding of a pollution standard as more offensive. And, if the excess is not deliberate, allowing defenses of due diligence, or reducing the penalty accordingly, can deal with more trivial technical breaches. Saying that individual license standards tailor crime to the individual is not a very convincing objection if to do this is fairer to the individual operator and more realistic in terms of environmental impact. Many people object to bargaining not merely because the resultant standards seem too weak but because the enforcement of these standards seems to be bargained away too. If we accept the inevitability of bargaining, we must avoid this problem.

Given the uncertainty of future ecological impacts, however, there may be circumstances in which license standards should later be relaxed. Similarly the development of future projects in the same air or watershed should be allowed for and may indicate an overly vigorous standard for the first plant. But the higher performance abatement technology is usually cheaper to install from the beginning than to retrofit, so in the long run the first plant of a series will save money. We are stuck with some sort of bargaining because the legal requirements of regulation prove to be an impediment to the proper use of knowledge. Is the issue one of the onus of proof—of proving cause and effect, of proving that harm would happen in the future—or one of using statistics to prove that cancer was probably caused by the defendant's chemicals? If these are the sorts of things Professor Thompson was thinking of, they all come back to problems of uncertainty and not of the law's unreasonable requirements. Surely a defendant has the right to insist that cause and effect are demonstrated before liability is imposed. On the other hand, if Thompson meant to say that the law simply asks the wrong sort of question, perhaps he is suggesting an expansion of the law's reach into areas where it is not competent. The court's desirable imposition of procedural fairness on a hearing process is a far cry from its deciding, for example, that the hearing panel, after balancing all the factors, was wrong in approving a nuclear power plant. Neither Professor Thompson nor I would wish such important decisions to be lifted out of the sometimes irritating but nevertheless theoretically accountable political process.

In posing deep questions about how sound public policy decisions can be made with limited time and information, Professor Thompson is really stimulating speculation far beyond his main subject. His paper really raises the question of whether modern states are governable in a time of almost overwhelming rapid change. Are our institutions—beyond the legal ones—outmoded? Journalists, academics, and Members of Parliament alike have observed a steady deterioration in the power of legislative bodies to scrutinize complex policy matters and financial estimates in a meaningful way (D'Aquino et al., 1979; Royal Commission on Financial Management and Accountability, 1979). Honorable Robert Stanfield (1978) has reflected thoughtfully on these ominous trends. As well as those matters that come before Parliament, consider the mind-boggling mass of regulation, policy directives, and other exercises of discretion in the bowels of departments or agencies. In all of these areas, whether legislative or not, policy directions set in earlier and simpler times tend to imprison us in "faits accomplis." For example, Canada committed itself to the development of nuclear power a generation ago, in the absence of a great deal of later scientific and social information. But when was the last meaningful parliamentary debate about the scientific and moral questions surrounding nuclear power? Perhaps admitting the impracticality (or relevancy) of that forum, most jurisdictions have tended to carry on internal reviews (the federal government of Canada), or to name a special parliamentary committee (Ontario), or to allow so-called peripheral questions about the whole nuclear cycle to be raised willy-nilly in the context of a particular project assessment (the El Dorado proposals for nuclear fuel processing under the Federal Environmental Assessment Review Process).

One of the most vexing problems about nuclear power is the need for extraordinary security to prevent spent fuel from being used by terrorists. Sir Brian Flowers, chairman of the United Kingdom's Royal Commission on Environmental Pollution, said in 1976, "I do not believe it is a question of terror or blackmail, but only of when and how often."* The response to this security threat in the United Kingdom was the Atomic Energy Authority (Special Constables) Act 1976 which set up a paramililtary, armed police force under the Central Electricity Generating Board, which reports to Parliament through the Secretary of State for Energy, who is not answerable to Parliament for their day-to-day activities.

Here is an armed police force, uncontrollable by the law officers of the Crown (and probably by the Energy Secretary), in a country whose regular police are unarmed. Its standing orders will not be published,

*Speaking at National Energy Conference, 22 June 1976, quoted in Flood and Grove-White (1976: 6n).

and the parliamentary debate was marred by the inability of M.P.'s to obtain information on what, if any, security measures were already in force. It was apparently felt that political control could limit the force's flexibility and secrecy and hence its effective operation.

Flood and Grove-White (1976: 20), after listing the threats which the powers needed by intelligence and police forces posed to civil liberties, concluded "the brutal message of the . . . Act seems to be that plutonium security is not simply more important than democratic controls. It may actually be *incompatible* with those controls."

How is one to respond? In Flood and Grove-White (1976: 30), one man said: "We are going to need a mass movement on the scale of CND [Committee on Nuclear Disarmament] if we are to stop this nuclear madness. . . . We are going to be forced into being outrageous." And who is this revolutionary Marxist who has been, no doubt, marked as a potential subversive by MI5? It is Lord Avebury, then president of the Conservation Society.

After this chilling (and perhaps extreme) example, it is almost anticlimactic to learn that public participation in planning approvals and the siting process for nuclear plants will probably be overriden in the future. But think of the implications: A closed decision-making system, with a deeply polarized population, wherein people with reasoned cases will not even be heard. Will not violent dissent follow? If there are many large-scale types of projects with potentially catastrophic results, can we really hope that a more open bargaining process will allow these deep value disputes to be satisfactorily adjudicated? Can our huge bureaucratic institutions handle the job? Perhaps we are in a period, like the Renaissance, of deep-seated cultural change (Nelson, 1980). Must we then examine our society to see if fundamental reorganization of our decision-making and production machinery is needed? For example, should we decentralize decision-making to the maximum feasible extent?

Speculations of possible new directions are a dollar a dozen (inflation you know) but we should perhaps recall certain assumptions which our capitalist theory still makes and which, together with other materialist political or economic philosophies, militate against improved environmental quality. Competitive individualism, adjudicated by a self-regulating market, will result in an efficient allocation of resources and even a just one if certain assumptions are made:

1. No one actor is big enough to affect supply or demand and hence price.
2. There is perfect information.
3. No externalities exist.
4. Perfect mobility of the means of production (including people) has been achieved.
5. If just acquisitions and transfers occur, that is, if there is no exploitation, the end result will be just no matter how rich or poor people may become.

This just, efficient resource allocation is supposed to happen through the invisible hand of enlightened self-interest which will maximize individual and societal well-being. But these conditions do not pertain, and the invisible hand is unhappily at our throat. Not all capitalists were moral, exploitation ran rampant, and governments had to intervene. Private profit did not equate with social benefit. What this means in terms of environmental regulation is that there are powerful economic concerns who profit by minimizing production costs, such as pollution control equipment. It is another example of the Tragedy of the Commons (Hardin, 1968) or the free rider (Olsen, 1965, 1971) that it is always in your individual interest (since everyone is doing it) to use the commons (air and water) for your profit.

If we are forced, as Andrew Thompson persuasively argues, to use a bargaining approach to environmental control, what do we face? A sophisticated, wealthy bargainer whose interest is the status quo, one who often has a monopoly on detailed financial and process data, so that we almost have to take its figures as correct. Recall the fluctuation in Canada's oil and gas reserve calculations by the National Energy Board which largely accepts industry's data.

Across the table sit civil servants, usually highly competent but with unclear mandates, and quite possibly untrained in weighing the public interest either ethically or in sophisticated public policy terms. They accept the agenda of questions which industry wishes to debate and, except through their minister, are neither visible to the public nor accountable to it. But they are in constant contact with the industry being regulated, and the odds are that they have the same education and value system. They may even hope to be hired away from government by the industry at a substantial increase in salary.

Of course, we will expect industry to win this bargaining game, especially when the general public is hypnotized by continual claims that environmental controls add to delay, cost of products, and hence unemployment. Perhaps bargaining need not inevitably entail these sad results, even if it always has to this point. Perhaps admitting well-funded interest groups into the bargaining process (procedural fairness) could help. Perhaps a more aroused public would in turn spur governments into more rigorous negotiation. Perhaps today we are merely getting the level of environmental quality we deserve. For the public gives a two-sided message. Public opinion polls show that we as a society would be willing to pay higher taxes and prices for a cleaner environment. Yet we demand a level of material consumption that renders environmental degradation inevitable, and we complain when taxes or prices really do go up. And we blame the government.

One of the most difficult problems we face is the belief of many citizens that they have the right to their particular subsidy—that is justice—and to protection by the government from the slings and arrows of rapid change. Yet they become alienated from their government when it helps other people with subsidies—that is socialism. They want lean, mean government for everybody but themselves. Unless we wake up and reestablish feelings of community, cooperation, and responsibility, our dog eat dog world will become an increasingly bleak, inhumane, and violent place. Not all the bargaining, not all the criminal law can hold together a society whose members do not like or trust each other.

We may conclude with some simple nostrums. As a society, we must give serious attention to training our citizens in ethical analysis. Much environmental harm results when one group of people solves its problems at another's expense without that other group's consent or without compensation. No bargaining model should ever countenance this situation. Obviously, therefore, attempts at social impact analysis must identify who benefits and loses by various possible alternatives. We will also have to continue making some decisions without adequate information. Unhappily, uncertainty will continue to be a factor in our complex world. But this does not doom us to flipping coins, or to the uninformed stupidity of treating all decisions as merely value questions and not as technical questions at all. Values are necessary in making social judgments, but if scientific or technical matters pertain, values will not be sufficient. Correspondingly technical matters are almost never devoid of ethical and social implications.

Perhaps in light of confusion and uncertain knowledge, we have to say that the proponent has not met his or her burden of proof which would justify the project. Perhaps postponement is the prudent decision. And, if some proposals are incompatible with democratic controls, it seems clear that they should not be countenanced by democratic societies. Perhaps, too, we have to downscale our projects. If serious harm can result from large-scale activity, why not simply use pilot projects and increase their number or scale as we learn from experience? Thus, if disaster strikes, it will be much less harmful.

The real question we are left with is how to restructure our political processes to make them more accountable and to allow them to use the best social, philosophical, and technical thinking available in a time when there are serious conflicts of values. Certainly open public hearings and freedom of information will be essential parts of any restructuring. Furthermore, decision-making mandates must be broad enough to include social and ethical considerations. It is inconceivable, for example, that a series of incremental decisions should be allowed to lock us into such an unforgiving technology as nuclear power without the fullest opportunity to examine its implications. Serious proposals (Vickers, 1970; Robertson, 1978; Henderson 1978) for decentralized, smaller, more manageable institutions have been made. As powerful centripetal forces work in the opposite direction, we will have to design them self-consciously, rather than muddling through. In the energy field Hooker et al. (1981) have made many extremely useful suggestions.

These ideas are especially apposite in a country like Canada, where strong divergencies of regional interests exist and attachment to central institutions seems to be waning. (This is true at the provincial as well as the federal level.) It would be tragic if we failed to respond imaginatively and allowed our political structures to collapse through successive bouts of inattention.

5

Counterpoint: Sources of Uncertainty

JOSEPH DISANTO

Professor Thompson's first point concerned the limitations of science and technology in the context of government regulations. He pointed out that the expectations of science are relatively high and the use of science in decision-making is relatively low, and with this I concur.

I had the opportunity to review the Alberta Oil Sands Environmental Research Program (AOSERP, 1980). The first thing I noted was that no one had asked the decision-makers about the kind of information they might need to make decisions regarding the AOSERP program. Why does this occur? Decision-makers have overly high expectations of science, thinking that if they leave the scientist alone, he or she will come up with the answers to all the problems with which the decision-makers might be confronted. This is not the case at all. Scientists do not have a corner on the market in terms of knowing the kinds of information the decision-makers are going to require in determining whether resource developments ought to occur and the direction they ought to take. Scientists are not in the position to know because other information besides the scientific always goes into the decision-making process, including political factors, tradeoffs, and value judgments.

Environmental questions are not simply scientific ones; rather, they are questions of the ecological interplay between physical, biological, and social systems. We simply do not know enough about the *interactions* between these areas. This came through clearly in the AOSERP work when it was decided that, instead of a theoretical conceptual model to integrate these areas, an administrative model would be set up. The administrative model merely divided up the work among the respective scientific workers. At the end of several million dollars of research the question was how were the findings of these disparate studies to be integrated. Without any kind of clear theoretical conceptual model, integration after the fact became impossible because the numerous studies under each of the administrative systems were spinning off in different directions. There was no real unifying conceptual model to provide the basis for integrated decision-making.

There are reasons why the nature of the interactions between the various systems referred to is not known, and one of the most important is the way we train people. We train people, even within these various areas, as specialists. Anyone who has ever attempted interdisciplinary research knows that one of the major problems to be faced is that different specialists cannot talk to one another. There is no common language to help us connect the physical, social, and biological spheres. It generally takes a year or two to come up with a common language or jargon.

Various kinds of institutions have been developed to deal with the problem. For example, the faculties of environmental studies are supposed to be interdisciplinary. These faculties are made up of specialists, each protecting his or her own special interests. Often there is insufficient communication among the members representing diverse interests, though ideally it is supposed to occur. Furthermore, the students who graduate from these faculties want to develop certain kinds of marketable specialties and therefore tend to specialize within these interdisciplinary faculties.

Another problem in the area of the limitations of science and decision-making is that of institutional funding. First, in Canada natural and physical science is funded by the National Research Council (NRC), while social and political science is funded by the Social Science and Humanities Research Council (SSHRC). If you develop an environmental research project, based on interdisciplinary research, the NRC is likely to deflect it to the SSHRC, which in turn will refer you back to the NRC. To make matters worse, each of these councils is composed of specialists who, during the review process, stress the "need" for research support going for their own special area. Consequently an institutional resistance to interdisciplinary research has developed which might otherwise provide the basis for an integrated kind of conceptual approach required for environmental research. This contributes to the state of uncertainty that Professor Thompson mentioned.

In addition to uncertainty, there is the problem of value judgments; in fact, value judgments may pose more of a problem than uncertainty. In a 1980 hearing of the Energy Resources Conservation Board on the extension of the Highvale Surface coal mine in Alberta, two experts dealing with the problem of noise pollution were in almost total agreement about decibel levels produced; there was no uncertainty. The noise decibel level produced by a particular operation would be above the ambient level in the area. But the value judgments they could not assess involved the source of the noise. There is "good" noise and there is "bad" noise. One of the uses of this particular geographic area is snowmobile recreation. Is there anything noisier than riding a snowmobile? Is it much noisier than the heavy mining equipment that is already in operation down the road? The government regulator is put in the undesirable position of having to ascertain not only the actual level of noise, which is a scientific matter, but also whether the source of the noise is more or less desirable and from whose perspective. He or she has to make a *social interpretation* of the noise.

The same problem occurs with dust in a related project involving the opening of a strip coal mining operation. In discussing relocation with local farmers and ranchers, they were told that the mining operation was going to be very dusty and that it would be extremely unpleasant. But these were farmers and ranchers who work around dust all the time, such as with a combine operation. But that is "good" dust. Coal dust by contrast is "bad" dust. So again the question was not simply one of how much dust was produced: the scientific evidence could spell this out quite reliably. The real question was the social interpretation of the source of the dust. There is good dust and bad dust, depending on one's point of view. It follows that in many of the areas of pollution and contaminants the issue is really more a question of value judgment than of uncertainty.

A similar point was raised at a public disclosure for siting a thermal power plant at Genesee, Alberta (1978). The architect had proposed a beautiful architectural structure superimposed on this pastoral scene. For the architect it was a work of art, but coming from the perspective of a farmer, it was the ugliest atrocity that could ever be imposed on that particular environment. Again, the question of aesthetics emerges over and above the simple scientific questions.

Several years ago an article appeared on the sociology of odor (Largey and Watson, 1972: 1021–34). If one does not think there can be a sociology of odor, all one has to do is look at the city of Hinton, where social status is reflected in how far away from Hinton's pulp-producing plant people live. People from outside the community claim that one cannot even get rid of that odor, so that when a resident of Hinton walks into Jasper or some of the surrounding communities, the odor is still in one's clothes, body and hair. That person is looked down on because of the odor of the pulp plant in Hinton. Again, in addition to questions of uncertainty in science, it is this kind of value judgment or aesthetic judgment that we have to be a bit more concerned about. Hence, while I agree with what Professor Thompson says about uncertainty, it is clear that the uncertainty is not just uncertainty in science. It is also a question of values and aesthetics. Such matters as the sociology of noise or the sociology of dust remain to be studied. These questions might really be considered red herrings. We have very little knowledge of how such matters operate in the society. This is a blind spot in sociology.

Values also come into play in the application of prohibitions and penalties, especially in the area of pollution. Aside from the legal uncertainties in going the legal routes to curtail pollution, we must keep in mind the relationship which pollution—or rather waste—has to our whole way of life. According to the second law of thermodynamics, as the scope of technology increases and as we mobilize more materials and power from the environment, pollution is inevitable (Schnaiberg, 1980: ch. 1).

Now, imagine the dilemma of the regulator who has to choose between pollution or economic depression. If we reduce pollution, then we have to reduce the amount of energy and matter that we are extracting from the environment. What does this do to societal values? It means that we do not have the consumption levels we like, and we do not have the lifestyle we prefer. If we cut back on pollution, we have to cut back on our material standard of living. Consequently, dealing seriously with pollution has implications for the magnitude and style of economic development. It implicates the rate of resource mobilization for producing goods. Therefore, legislation with penalities to control pollution must be organized with an eye to its potential effects on the economy.

We must consider who wins and loses with this kind of problem. This raises the issue of who bears the costs of such legislation, which brings us to Professor Thompson's last point, the role of public participation, an area that is quite complex as has been demonstrated in my own involvement with public participation in resource decisions.

By public we usually mean the people who will be most directly and/or adversely affected by a project, the people who are most clearly acknowledged in the decisions of the regulator agencies. The general public or public interest is often not very well defined and not very well developed in decision making. What is the public? What publics are the regulatory agencies attempting to protect? The agencies must very quickly partition the public into sectors to acknowledge competing interests and competing ideas of the costs and benefits of pollution and production. When we introduce the cost-benefit analysis, we must differentiate between various kinds of publics, and we discover that what is a cost to one group is a benefit to another. As a consequence, this raises questions of equity, a theme that runs throughout papers in this volume.

As ecologists and sociologists, we are often concerned about the notion of equity. But what is so sacred about the notion of equity? Ecological systems are arranged in terms of ecological niches. Different species, different groups, and different activities are organized in a way that they fit into various niches. The relationships between these niches are relationships of power, subservience, symbiosis, and so on. Hence, some niches get more than other niches. Equity is often interpreted to mean equality, or in this context equal treatment or "equality of condition" between different niches. But this is not the only way to view the issue of equity, and it is not the way ecological systems normally operate. In them, conditions are unequal. It follows then that if we are really to deal with the problem of equity, rather than redistributing energy in an attempt to make conditions equal, what we really ought to be concerned about is redistributing the niches or opportunities and positions in the system. Is this not in fact the way society operates? New niches are being formed in society all the time. For example, the computer explosion brought new occupational niches into society. New occupations came into society, and occupations disappear from society as they do in any other ecosystem. It is not surprising that there are differences in equity, in equality, in cost, and differences due to some groups having to bear the cost of pollution and some groups reaping the rewards, because that is the way all systems seem to be organized. Perhaps the solution to the problem of equity is to enhance the ability of groups and individuals to move between different niches, rather than to attempt to equalize the conditions among them.

6

Technology, Safety, and Law: The Case of the Offshore Oil Industry

W. G. CARSON

With the loss of the semi-submersible drilling rig *Ocean Ranger* and its crew of eighty-four men in the North Atlantic off the coast of Newfoundland in February 1982, a whole series of issues relating to the safety of offshore oil operations has once again become the subject of heated debate. First, of course, there is the still open question of what actually went wrong, a question that has generated a profusion of speculation, although the final answer will only be known when the ruptured structure has been examined and the results adjudicated by the different inquiries set up to investigate the disaster. Inevitably, too, considerable media attention has been focused on the adequacy of the safety standards and procedures followed by the rig's designers, constructors, owners, and operators, and there has been no dearth of suggestions that at various stages along the line all was not quite as it should have been in this respect. Moreover, this disaster has raised a number of important and even embarrassing questions for the relevant Canadian authorities with regard to the adequacy of their provisions for ensuring the safe operation of installations working off the coast of Newfoundland. Suggestions that certification may not only have been in the hands of U.S. bodies but may also even have been out of date have done little to enhance the credibility of either federal or provincial governments in this area, while the continuing jurisdictional dispute between St. John's and Ottawa has been given a new and nasty twist by the catastrophe. Lurking not far beneath the surface of public debate, moreover, is a sense of surprise, or even of belated outrage, that op erations could have gotten underway in the absence of an adequate legal regime for ensuring the safety of operations.

In this paper it would obviously be inappropriate to preempt official inquiries by pursuing discussion of the *Ocean Ranger* disaster in any detail. Many of the issues raised, however, are highly reminiscent of questions that have plagued the much longer-running experience of the offshore operations that have been taking place in the British sector of the North Sea since the early 1960s. While that particular location has not thus far witnessed any multi-death accident to compare with the tragedy off Newfoundland, the record in terms of less spectacular deaths and injuries has certainly not been a good one, while there is some evidence to suggest that, at least on occasion, outright disaster has not been all that far away. Nor have industry and the British authorities lacked for critics of their respective approaches to safety. Over the past fifteen years or so, there has been a sporadic barrage of allegations from union leaders and others to the effect that, despite its protestations, the offshore oil industry puts profit before safety. Particularly during the early stages of development, the authorities displayed scarcely any more prescience than their Canadian counterparts in recognizing the need to institute adequate legal safeguards for the protection of those employed on the Continental Shelf. Indeed, the whole history of the official response to the safety implications of exploiting Britain's offshore oil and gas resources has arguably been one of safety taking second place to the perceived urgency of getting those resources ashore (Carson, 1981).

One theme running throughout the debates surrounding Britain's North Sea success story has been that of the difficulties inherent in utilizing legal controls to regulate an enterprise dependent on technological innovation. At one dangerous extreme, it has sometimes been suggested that there is little the law can do when the industry in question is operating at the very frontiers of technology, not to mention under extremely adverse operating conditions (Kitchen, 1977: 129). More generally, the development of legal responses to the question of offshore safety has constantly been stalked by the spectre of statutory intervention hampering the very technological developments on which rapid exploitation of the Continental Shelf's resources was perceived to depend. Concomitantly, the authorities' disadvantaged position vis-à-vis the technical expertise and information available to the industry has been explicitly acknowledged as both a major difficulty confronting the evolution of legal controls and as something that necessitated highly specialized institutional structures for the formulation and implementation of such controls. Thus, although technological issues cannot be said to have been the central driving force behind North Sea developments, they have beyond doubt played an important part in shaping the legal response to the evident occupational hazards involved. It is this role that I wish to examine in the remainder of this paper.

TECHNOLOGICAL FRONTIERS AND THE LIMITS OF LEGAL EFFECTIVENESS

In the wake of the *Ocean Ranger* disaster, one of the most powerful images of the offshore oil industry rapidly came to the fore—that of an industry operating at the very frontiers of technology in its battle with extremely adverse, if not always horrific, climatic conditions. According to the general manager of the Canadian Association of Oil Well Drilling Contractors, for example, it was appropriate to invoke the unpredictable force of the elements in connection with the accident. Such rigs, he explained, "are designed to withstand storms (but) a lot of the time we don't understand mother nature and what she can do" (*Toronto Star*, 16 February 1982). While such a comment scarcely reflects favorably on the record of the research that has been going on for some years in the not dissimilar setting of the North Sea, it accurately captures one side of the technology/natural elements imagery that has become such a salient feature of how offshore operations in northern waters have been popularly represented over the years. As for the other side, there seems to have been some feeling that the disaster indicated that technological frontiers might not only have been reached, but also transgressed. The point was put bluntly by one well-known Canadian radio commentator in summing up what he believed was a widespread feeling: "Since the disaster of the *Ocean Ranger*, many people have been wondering whether or not the elements aren't just too much for man to overcome. Maybe we would be better off to just stay away from drilling in such rugged parts of the world" (Canadian Broadcasting Corporation, 19 February 1982).

Implicit in such a statement is the suggestion that in conducting oil operations in settings like the North Atlantic, the current state of the relevant technology may, at least temporarily, have met its match. Thus, we either surrender to superior natural odds, or, more sanguinely, we accept the fact that tragedies of this order will be an inevitable, if hopefully not too frequent, price that will have to be paid for recovery of the mineral resources lying beneath the seabed. Although no one was sufficiently insensitive to give explicit expression to the latter view, some observers came close. Thus, for example, shortly after the accident took place, one Ottawa commentator voiced his fears that Newfoundland's newfound hopes of economic prosperity might be dashed if the tragedy led to suspension of operations, even though, as in other industries, major disasters do occur. According to another reported comment on the possible consequences of the tragedy, the relevant technology would certainly be questioned carefully "but the need for the oil off Canada's east coast is too great to halt the development" (*Calgary Herald*, 20 February 1982). Similarly, and in terms at least by analogy conducive to reconciliation to the inevitability of disaster, one of Canada's leading newspapers quoted Newfoundland's Energy Minister as being careful "to note that, although tragic, the *Ocean Ranger* sinking is no different from other marine disasters to which Newfoundlanders have become inured over the years" (*Globe and Mail*, 17 February 1982).

In more fully developed form, a similar imagery of frontier technology, of adverse operating conditions, and even of death and injury as an inevitable price that has to be paid has also been a prominent feature of how journalists, politicians, and others represent the risks associated with offshore activities in the British sector of the North Sea. As Guy Arnold (1978: 9) remarks, for example, "North Sea oil has created its own language: the leading phrase is 'on the frontiers of technology' and everyone loves the image this creates." Similarly even the industry's own project engineers are said to have been left breathless by the pace of technological developments that have been stimulated, not just by the need for change in itself, but by "the peculiar circumstances that engineers have found themselves dealing with in the North Sea" (*Scotsman*, 3 September 1979). Nor is there any dearth of reports that link the challenge of the offshore enterprise in these respects to its implications for safety:

> The North Sea has presented man with one of his biggest challenges this century . . . a wealth of energy lying 12,000 feet below turbulent 600 feet deep seas. An area which can produce winds up to 160 mph and waves of 100 feet—as tall as an eight or nine-storey building. It is here that the offshore industry has constructed the necessary platform giants to extract oil and gas and send it

ashore. . . . In such a hostile area, safety and the environment provide the starting point for all decision-making. (*Aberdeen Press and Journal*, 10 August 1979)

Whatever the priority allocated to safety in the planning and execution of offshore operations, the frontier image is one that readily reconciles readers to the inevitablity of accidents. People are killed at inhospitable frontiers. Thus, a further image that is often projected in discussions of offshore safety is that of necessary sacrifice for the common weal. While we may dismiss as mere, if possibly inappropriate, humor, the view of one Treasury man who explained that, "with the economy in the state that it was, the people who were dying there were dying for the greater good,"[1], the suggestion of necessary sacrifice is often quite explicit and quite serious. Thus, for example, in 1977 Anthony Wedgwood Benn, the then Secretary of State for Energy, wrote about the reduction of offshore risks. Having outlined the vital contribution oil and gas could make to Britain's future, he turned his attention to "the penalities" which "as with all things . . . are to be paid." In terms possibly more redolent of the cenotaph than of the North Sea he went on: "Too many have already paid the ultimate penality with their lives, which is tragically the price so often extracted of pioneers" (*Guardian*, 21 June 1977). More recently, the *Guardian* deployed the metaphor of cost-benefit analysis to provide a stark caption for a discussion of "the human price of Britain's oil billions" (*Guardian*, 10 August 1977). This and several of the other images already mentioned were cogently combined in one editorial in the *Aberdeen Press and Journal* (10 August 1979):

North Sea oil and gas will be worth a staggering 7,200 million pounds this year—and we owe an enormous debt to all who have helped to bring this about. The costs have been high and grievously so in regard to the loss of life which has been incurred. Only yesterday we had another grim reminder of the human toll involved in this vast and crucial operation, in often severely testing conditions which demand taking technology to its outer limits.

Not all of the images associated with offshore employment can be dismissed as false. More specifically there is little doubt that, at least until very recently, the North Sea has been a very dangerous place indeed to work. While the "human toll" of over one hundred killed and more than four hundred seriously injured may not seem particularly appalling in absolute terms,[2] the incidence rates lying behind these stark statistics compare very unfavorably with the rates in other reputedly dangerous occupations. Thus, for example, between 1974 and 1976 the overall risk of being killed on or around an oil installation in the British sector (excluding accidents on vessels, for which no employment figures are available) was around six times that for the quarrying industry, nine times that for the mining industry, and eleven times as great as in the course of construction work (see Carson, 1981: ch. 2). When diving, probably the most hazardous civilian occupation in contemporary Britian, is left out of the account, the gap narrows (to four times, six times, and eight times, respectively) but by no means disappears; when the figures relating to diving itself are calculated over the same period, the resulting incidence rate for fatalities is in the region of eight per thousand or twenty-six times the death rate in quarrying, thirty-eight times in mining, and fifty times that in the construction industry. While the record for diving, as for the industry as a whole, showed substantial improvement toward the end of the decade,[3] there is no doubt that the North Sea thoroughly deserved its dangerous reputation during the most crucial phase of its development in the mid-1970s.

If such is the case, it becomes germane to ask why offshore employment on the British Continental Shelf has been so hazardous? This is a question to which some of the other images already mentioned have considerable relevance since they suggest, at least by implication, that the high casualty rate has indeed been an inevitable consequence of working at technological frontiers and in consistently adverse climatic conditions. No less important, they imply that there is relatively little that the law could have done, or can do, to minimize the human cost exacted by the unique exigencies of the offshore situation. The other price of Britain's oil, they suggest, is not negotiable.

Elsewhere it is argued at some length that, although there is no denying the industry's technological sophistication or the harshness of the setting in which its operations take place, neither of these factors, so salient in the imagery woven around the North Sea oil industry, can be held primarily accountable for the rate of death and injury that has occurred (Carson, 1981: ch. 3). By using data taken from fatal and serious accident files, it was possible to show that the vast majority of the accidents in question emanated from relatively mundane causes comprehensible to anyone familiar with the field of industrial safety in general. Unsafe working practices, poor design and maintenance, inadequate communication and supervision, shortcuts taken under pressure, and lack of elementary safety precautions were all shown to have played a depressingly familiar role in the genesis of offshore accidents. Accordingly it was concluded that we should be wary of the superficially attractive explanation of high accident rates in terms of technological frontiers and adverse operating conditions. *Mutatis mutandis*, I suggested that many of the risks that have so tragically materialized in practice are potentially amenable to regulation by law, even if the legal response from the relevant British authorities over the past fifteen years or so was very slow to take advantage of that potentiality. To be sure, this does not mean that I am idealistic to the point of asserting that the law can ever ensure a totally safe working environment; there is too much empirical and theoretical work suggesting the contrary for such a view to be seriously entertained. All I do maintain is that there exist no *a priori* grounds, least of all those based on the imagery of frontier technology and adverse operating conditions, for simply assuming that the law must stand helpless before the catalogue of death and injury which has been compiled in the course of British offshore operations to date.

One possible objection to the foregoing argument is that, while it may be valid for the British context where the accident record is largely one of workers being "picked off" singly or in pairs rather than as a result of major catastrophe,[4] its applicability to a disaster such as that which overwhelmed the *Ocean Ranger* is both undemonstrated and more dubious. Surely, it might be said, when a structure apparently claimed to be unsinkable and standing as tall as a multistory building keels over in a storm, this could only have been the result of a technology that trespassed beyond the limits of its own capacity to handle the elements, a sort of fatal frontier incursion.

In the absence of any definitive explanation of what actually happened on this occasion, such a possibility must, of course, be left open, though reports that nothing more technologically sophisticated than the smashing of a porthole in the ballast control room precipitated the disaster scarely bode well for the frontier-type argument. Moreover, the only comparable precedent—the capsize of the *Alexander Kielland* in the Norwegian sector of the North Sea in March 1980, with the loss of 123 lives, similarly counsels a degree of caution. According to the official report of the inquiry into this accident, the immediate cause was fairly evident (Utredninger, 1981). Into an opening cut in one of the six bracings holding a column (leg) to the platform a support for a hydrophone positioning control had been welded. This welding was "not the best," and the fact that the hydrophone support would now comprise part of the load-carrying structure, as opposed to "outfit" (equipment), had not been considered at the planning stage. Fatigue fractures had developed in both the welds connecting the support to the bracing and in the support itself. Some of these fractures must have developed before the platform was placed in position because, among other things, paint residues from the construction yard where the installation had been built were subsequently found on fracture surfaces. Neither control over design planning nor surveys carried out during building and operation had been sufficient to reveal these defects.

On the day in question, the weather had been bad, though not, it may be added, particularly ferocious by North Sea standards. Wind velocity had been 35 to 45 miles per hour, and wave height 20 to 25 feet. By this stage the fatigue fractures already mentioned had produced a redistribution of stress, initiating further fatigue cracks in the bracing to the point where two-thirds of its circumference was affected. When the weakened bracing broke, its five companions were subjected to overloading and gave way in rapid succession. The column became detached from the platform, and the *Alexander Kielland* heeled over to an angle of 30 to 35 degrees, largely because neither design nor regulations had allowed for the need to maintain stability in the event one of the five columns broke loose. Even then, disaster might have been averted if there had been compliance with instructions pertaining to watertight doors and ventilators on the deck. However, with some possible assistance from holes caused by damage already sustained, these openings allowed over 50 percent of the deck volume to flood, and the whole structure turned over completely in the space of some twenty minutes. This drastically reduced the time available for any attempt at rescue from the installation itself, and apart from those who remained to go down with the rig, the rest had to face the sea. As the report concluded, "The chance of surviving more than half an hour in the cold water was, as it turned out to be, minimal." Out of a total complement of 212 only 89 survived.

In all of this, there is very little to contradict my central thesis about offshore accidents. To be sure, the scale of the disaster outstrips by a long way anything that has thus far taken place in the British sector, but the chain of precipitating factors is depressingly familiar. While underwater inspection to detect weld failures or cracks may indeed be extremely difficult, technological ingenuity is scarcely taxed if, as would seem to have happened in the present instance, they are there to be found at the pre-installation stage.[5] Similarly, failure to allow for the loss of the crucial buoyancy column must surely strike even the layman as analogous to designing an aircraft without considering the implications of the failure of one of the engines. One expert who commented on the safety of Pentagone rigs like the *Alexander Kielland* in the wake of the disaster used the same analogy to even more powerful effect when he observed that, although the design was basically good, if it was not made structurally sound, the consequence was like "building an airplane without making sure the wings stay on" (*Sunday Times*, 5 April 1981). Failure to close watertight doors speaks for itself.[6]

Clearly, then, we must be careful not to be misled by the sheer scale of disaster or by the apparent complexities of its provenance. The factors leading to the *Alexander Kielland* disaster were knowable; their remedy lay within the range of existing technology; and they were almost certainly avoidable if a properly constructed, efficiently administered, and faithfully adhered to system of legal regulations had been applied. Moreover, the British authorities' complacent belief in the superiority of their own regulatory approach will retain even the semblance of credibility only as long as a similar tragedy does not overtake an installation operating in the British sector of the North Sea. The fact that a major installation working in that location had to be evacuated and towed ashore because of cracks that had appeared in its structure shortly before Christmas 1981 suggests that such an eventuality is by no means beyond the realms of possibility (*Glasgow Herald*, 12 December 1981). Indeed, just such a possibility had been explicitly acknowledged some months earlier when a Committee of Enquiry into Offshore Safety (the Burgoyne Committee) had reported that "it is known that in a number of semi-submersibles, weld failures have been so extensive that the safe operation of the entire rig was in jeopardy" (Burgoyne Report, 1980: 114). Such a comment, together with a letter sent to the present writer by one who claims to have worked on the construction of the *Alexander Kielland* and who observed that work practices inconsistent with official standards were demanded, make somber reading and suggests that there may well be reason to allow the law a more extensive role in the prevention of major offshore disasters.[7]

THE SOCIAL CONTEXT OF TECHNOLOGY'S ROLE

Thus far, the argument advanced here may seem to amount to little more than a "debunking" exercise with regard to the technological sophistication of the offshore oil industry. As already suggested, however, it is no part of my purpose to engage in such an enterprise save insofar as technological frontiers per se are advanced as an explanation for ineluctably high casualty rates. Nor, having attempted to dispel that particular "myth," do I see this as an end to the relevance that technological issues may hold for the broader issue of how satisfactorily the safety of offshore operations has been legally regulated. On the contrary, in the remainder of this paper, I will attempt to show how questions involving technology, technical expertise, and informational gaps between governments and industry on this plane constituted one vital link in a causative chain that spawned an offshore safety regime falling far short of the ideal in both structure and operation. More specifically I will argue that the deference accorded to the industry's technology and expertise did indeed, at least on occasion, permit development to outstrip the known, although not necessarily the knowable—a high-risk strategy that may yet result in further catastrophe. It will also be argued that technical expertise was one of the trump cards played in a complex game surrounding the formulation of statutory controls, the construction of an institutional structure for overseeing the safety of offshore operations, and the subsequent development of a very special regulatory relationship between controllers and controlled. Paradoxically, perhaps, I will also draw the conclusion that technological issues even had a hand in creating a situation in which the more mundane and hitherto more consequential risks associated with North Sea operations came to be comparatively neglected by law.

Here, however, it should be stressed that, as stated above, these questions of technology, technical expertise, and so on, serve as a "link," albeit a crucial one, in a causative chain. Thus, while most of what is to follow will be devoted to their discernible effects on the British offshore safety regime, it is relevant to pause here for a moment in order to ask where this same chain terminates if pursued in the opposite direction. Less enigmatically it may be maintained that the role played by these factors was itself contingent on the broader material forces surrounding the development of Britain's offshore oil and gas reserves. In short, that discussion of the part played by technology in this specific empirical context, as of its wider role in contemporary society, cannot be divorced from the underpinning issues of political economy. It is only by making this kind of linkage that the personal troubles so tragically and so frequently encountered in the course of offshore employment can be reconnected to the public issues, to the historical transformations, and to the "big ups and downs" of the society in which they are endured (Mills, 1959: 226). That the line of reconnection runs through technological issues and arguments should not blind us to the fact that technology itself has to be located within the broader social context which determines the pace of its development and the pattern of its deployment (Marchak, 1979: ch. 2). Concomitantly the analytical issue for law is not simply one of whether legal regulation can control technology as a kind of spontaneous development or as an autonomous outcropping of human inventiveness. Rather, the question is how law, often through the mediating influence of technology and its associated ideology, is itself constrained by material forces emanating from political economy.

In this paper, considerations of space obviously preclude elaboration on this wider issue in any great detail. Stripped down to its essentials, however, the general argument is that the opening up of the British Continental Shelf to exploration and exploitation has to be set against the backdrop of developments within the world economic system and against the background of the political and economic events that have shaped the pattern of oil supply, consumption, and pricing since the Second World War. Together, these factors may be described as the external dynamic in a process sometimes referred to as "combined and uneven development."

On the other hand, and as this term itself implies, it is also important to realize that development, even within the world system of capitalism, is uneven and that movements within the international context cannot therefore tell the whole story with regard to Britain's oil. The nation-state, as David Purdy observes, should not been seen as "merely registering in its internal motions forces at work on a global scale" (1976: 271). In consequence, it also becomes imperative to locate analysis of the burgeoning offshore enterprise within some discussion of Britain's historically specific and even unique features during the operative period, features that provide one vital part of the key to the oil policies adopted by successive governments over the past twenty years. It is only when this interplay between external forces and internal exigencies is grasped that we can begin to understand what I have elsewhere called "the political economy of speed" (Carson, 1981: ch. 4). In other words, when the intricate intersection between developments within the world economy, its oil sector, and the unique discussions of Britain's predicament from the early 1960s onwards is mapped out, it becomes possible to identify a concatenation of forces that generated and sustained an overriding impetus toward the speedy development of the British Continental Shelf's petroleum resources.

One important consequence of this commitment to speed was accentuation of the extent to which the host state became dependent on the good offices of the international oil industry. To be sure, this relationship was not all a one-way affair—the companies also needed government in a number of vital respects, not least in that of creating the legal system of proprietary rights in offshore mineral resources without which exploration and production could not take place. On balance, however, and although the relationship did remain symbiotic, there is little doubt that the endorsement of speed as the number one priority did tip the balance of reciprocity fairly heavily in the industry's favor (Carson, 1981: 116ff.). Thus, for example, it was the need for haste that was consistently deployed to justify the extent of North Sea participation allowed to overseas companies. The argument was that British Petroleum and the partly British Shell, even with the assistance of the Gas Council and the National Coal Board, would clearly have been incapable of mounting such a massive operation on their own, or at least not at the pace deemed desirable. On the financial front too, the speed of development increased the degree of dependence on the industry and its financiers, since the task of meeting vast capital requirements over a comparatively short period came to hinge crucially on the capacity of major oil companies to finance ventures from their own internally generated funds (Carson, 1981: 123ff.). And where this source proved inadequate, there were always the international banks to fall back on. According to the Committee to Review the Functioning of Financial Institutions which reported on North Sea oil financing in 1978, over 60 percent of all the identifiable loan finance outstanding or committed by mid-1977 had been raised from overseas, particularly American banks (Committee to Review, 1978: 26). Thus, in addition to dependence on the capital invested by major oil companies from their own funds, the development of Britain's Continental Shelf came to lean heavily on what the above committee, with a nicely anthropomorphic touch, called "the contribution of non-residents to North Sea oil financing" (Committee to Review, 1978: 16). The point was put dramatically by one oil correspondent in 1978:

> If the UK had had to go it alone—in the way that British companies built the railway systems of the mid-nineteenth century in this country, which is after all a comparable venture—then the strain on the financial system might have been much greater. But the key to the financing of the North Sea has been the involvement from the beginning of the established international oil companies, who have been large enough to treat North Sea exploration and development as part of their world-wide activities. . . . Further, so dominant has been London as a centre of the international capital market, that UK banks, or international banks with branches operating aggressively through the international capital markets in London, have over the years developed a variety of techniques that have enabled them to put together financial packages to cope with the special demands of large-scale oil development. (*Oilman*, 18 November 1978)

But it was not only in terms of operational capacity and the industry's capacity to raise the requisite financing that the rapid development of Britain's offshore resources involved a marked degree of dependency, for the authorities were also heavily reliant on the industry in another crucial respect. Although one of the world's largest oil companies is nearly half British and other is wholly so, there is no doubt that when the mineral resources of the North Sea first came on the political agenda, British officials found themselves at a severe disadvantage with regard to information and expertise. Technological hegemony, of course, is one of the key factors in multinational survival, and the North Sea oil industry was to prove no exception, while on the less sophisticated plane that principally concerns us here, the potential of the United Kingdom's Continental Shelf was a largely unknown quantity. Whatever the industry may have known about its possibilities, and there have been some suggestions that this was more than it was telling at the time,[8] officialdom was working substantially in the dark.

Nor were the authorities particularly well placed to remedy this situation, at least to begin with. As the Department of Trade and Industry later recalled in discussing the predicament of its predecessor in this respect, the Ministry of Power could not even demand information on the costs or results of the operations that were taking place "as of right on the high seas" at the beginning of the 1960s. Indeed, the need to overcome this disadvantage was one of the primary bureaucratic motives for preparing the legislation that became the Continental Shelf Act of 1964 (Committee of Public Accounts, 1973: 123). Moreover, even when the gap was plugged in relation to "physical information" on results, serious doubts still remained as to the Ministry's capacity to evaluate what it was being told. A chronic lack of technical staff persisted up to and beyond the end of the decade, and seismic and drilling results were passed to the Institute of Geological Sciences for assessment. Subsequently described in less than flattering terms as "an academic institution with limited expertise in petroleum production" (Hamilton, 1978: 31), this body was grossly inadequately funded for the task in hand up to 1967 at the very earliest (Committee of Public Accounts, 1973: 77).

Part of the problem here was, of course, that the North Sea enterprise was new. But the difficulty was greatly exacerbated by the almost "Catch 22" situation created by an official policy that favored speed. On the one hand, successive governments wanted to find out as rapidly as possible what kind and scale of asset was not available to them, and for this purpose wished to encourage speedy exploration by the industry which alone could furnish the answer. On the other hand, the more successful they were in this endeavor, the less time available for gearing the administration up to the task of either handling the information or developing its own independent monitoring capacity. The evidence given to the 1972 Public Accounts Committee by the Permanent Secretary at the Department of Trade and Industry was given with precisely this dilemma in the background. Time and again, he reiterated his belief that the speed of exploration to date had been perfectly justified because of the need to know as quickly as possible about the nature of this valuable national asset (Committee of Public Accounts, 1978: 124). At the same time, he had to concede, among other things, that the problems confronting the Institute of Geological Sciences stemmed from the fact that "this was an operation which went off with tremendous pace and . . . that particular Office was not set up for it then" (Committee of Public Accounts, 1973: 77). In more general terms, his account of the authorities' capacity to handle the early stages of North Sea exploration provoked some blunt and almost rhetorical questioning from his examiners:

> Would it therefore be reasonably true to say that in contemplating the exploitation of this massive new piece of publicly owned area we did not take time to gear ourselves in terms of the information that our Government Department ought properly to acquire about the potentialities of the area in terms of funding and in terms of know-how available directly to the government? We were not geared to do it? (Committee of Public Accounts: 77–78)

By the time this question was posed, late in 1972, civil servants were able to claim that substantial progress had already been made on covering the information deficit. Justifiable as such claims may have been, however, the continuing difficulties in collecting and assessing data about North Sea operations should not be underestimated. As late as 1973, for example, information on costs as opposed to results was still being supplied on a voluntary basis, and even if some officials felt that this system operated fairly satisfactorily, they could not deny that it did nonetheless place them at a certain disadvantage (Committee of Public Accounts, 1973: 126). Nor would they perhaps have been quite as satisfied if they had been privy to the complaints of some of their colleagues who, around the same time, were apparently citing instances in which requests for information not statutorily required were being turned down (White et al., 1979: 40). Even after the emergence of a separate Department of Energy, doubts still remained with regard to the effectiveness of its "handling of drilling and other information given to it by the companies and the quality of input it, in turn, feeds into policy decisions" (*Financial Times*, 6 February 1975). While the British National Oil Corporation is said to have generated intense dislike among the companies precisely because it was "dispelling the mystique about their knowledge" (Arnold, 1978: 165), its own communication of general information and plans to the Energy Department has been described as "sporadic." In more general terms, as recently as the end of 1980, the Minister of State at the department was still lamenting its short supply of "experts who are trained in the relevant geological and geophysical sciences and in oil and gas engineering," personnel who were needed "to advise on oilfield developments and operations more generally."[9] As we will see, moreover, the way in which the shortfall in expertise was made up over the years was only to raise further questions about officialdom's dependence on the industry in this context. Above all, increased expertise and information flow did not resolve one perennial problem, namely, that governments intent on establishing the full scale of their offshore reserves as quickly as possible were still substantially dependent on the international oil industry to supply the answer.

For the present purposes, such summary and even cursory treatment of extremely complex issues must suffice. Nor is there room to elaborate on what by any standard has been a spectacular success story in terms of Britain's progress toward self-sufficiency in oil, a target that was reached during the summer of 1980, by which time some 80 million tons of indigenous oil were being produced annually, while natural gas from the North Sea was accounting for some 78 percent of the country's primary gas demand. Even less would it be appropriate to digress into an account of the almost inexorable process whereby Britain became hoist with her own oil petard, sterling's "silly elevation to petro-currency status" (*Guardian*, 4 June 1981) which seemed for a time to turn the North Sea oil into the manufacturing industry's executioner rather than its savior. The denouement, in terms of the current glut in world oil supplies, falling prices, and the consequent threat to oil revenues on which so much political store had been placed over the past decade, likewise falls beyond the scope of this paper. What does not, however, is discussion of the way in which the commitment to speedy development of the North Sea's mineral resources and the concomitant relationship of dependence which it spawned created a crucial role for technological issues in the shaping of legal responses to the question of offshore safety.

SPEED, TECHNOLOGY, AND STATUTORY CONTROL

The first and most obvious way in which the exigencies emanating from the political economy of the North Sea exerted an influence on the emerging safety regime was by the establishment of a set of priorities within which it was tacitly accepted that statutory regulation must not be allowed to stand in the way of rapid development. That such a hierarchy of priorities would be applied became evident from the very outset when, in late 1963 and early 1964, the creation of the legal framework required to get offshore exploration underway was under consideration by the British Parliament. Falling prey to what Petter Nore has dubbed "the fatal historical consequences of the British obsession with the balance of payments and the defence of the pound" (1976: 14), the Conservative government of the day made it clear that no delay would be brooked on account of the misgivings about safety which were already troubling some members. At one point in this not terribly creditable chapter of British legislative history, the Minister of Power even chided a Standing Committee of the House of Commons for the fact that the "biggest delay" was being occasioned by its deliberations and, in terms more redolent of an offshore toolpusher than a government minister, went on to explain that "every week counts."[10] In so doing, he was merely reiterating his government's avowed intention of pressing ahead as quickly as possible with the "highly desirable and urgent" objective of creating a system of proprietary rights in the mineral resources of the Continental Shelf, the *sine qua non* of facilitating speedy offshore exploration.[11] Time clearly was to be of the essence:

> This is not a matter of months. We know that the oil companies are extremely anxious to get going as quickly as possible. I emphasize the time factor in this matter. It is much easier to work in the North Sea in the summer than in the winter, and it is not a matter of waiting for months before anything happens.[12]

In such an atmosphere, it was highly unlikely that any doubts about the adequacy of arrangements for securing the safety of offshore operations would be allowed to impede the rapid provision of the legal regime necessary for further exploration. Nor were they. Despite repeated protests and acrimonious accusations that labor was not being accorded the same degree of protection as capital,[13] the government proceeded to do what it had all along intended. Instead of provision for statutory regulations, safety considerations were to be dealt with by means of "model clauses" inserted into exploration licenses,[14] while further safeguards would be provided by the extension of the civil and criminal law of Britain to the Continental Shelf.[15] Thus was created an extraordinary situation in which legal control over the safety of offshore operations would primarily take the form of contractual obligations, the main sanction being that of license revocation.[16] As for those who fondly hoped that extending the criminal and civil law to the Continental Shelf would place the industry under the control of land-based safety regulations, they were first to be confused and subsequently to be confounded: confused when a government spokesperson explained to some highly puzzled peers that "the Act which the Noble Lord mentioned, such as the Factories Acts and so on, where they apply, can be made to apply to an installation on the Shelf"[17]; confounded when the burden of legal opinion subsequently concluded that such installations could not be legally classified as factories (Kitchen, 1977: 204).

In all of this, technology as such did not figure very prominently, though throughout these early debates there were lingering suspicions that one reason for the government's reluctance to provide for the formulation of detailed safety regulations was its desire to avoid placing any obstacle in the path of the technological improvisation necessitated by a new and hostile geographical location. However, subsequent events were soon to show both the inadequacy of the system that had been set up and the explicity deferential attitude being adopted by the authorities toward the industry's technological status. In December 1965, only some eighteen months after the Continental Shelf Act had been passed, the jack-up rig *Sea Gem* collapsed and sank with the loss of thirteen lives, and the "voluntary" Tribunal of Inquiry which followed (voluntary because no provision had been made for statutory inquiry) quickly revealed some dire defects in the current safety regime (Report of Tribunal, 1967). The license issued to BP Petroleum Development Ltd. had indeed included the requisite "model clause," but this only obliged the company to abide by the provisions of the Institute of Petroleum's Model Code of Safe Practice—that is, the industry's own code (ibid.: 18–19). Moreover, the authors of this code themselves emphasized that it dealt only with recommendations, while its structure and layout were "not apt to make it clear and authoritative as a piece of quasi-legislation" (ibid.: 18). As for sanctions, the Tribunal spotted the regulatory problem right from the beginning. "It is perhaps as well to make it clear this early," said the introduction, "that the only sanction for ensuring the proper operation of the safety procedures is the revocation of the license" (ibid.: 2). Accordingly, the Report went on to include in its list of recommendations a crucial proposal for a *statutory* code supported by "credible sanctions" (ibid.: 24).

Five years after the *Sea Gem* collapsed, and some two and a half years after the report on the circumstances of that collapse was published, Parliament finally got around to the task of making clear statutory provision for the promulgation of offshore safety regulations. By then, oil as well as gas had been discovered in the North Sea; a Labour

Government had come and gone, thereby lending a bipartisan quality to the neglect of offshore safety; fourteen more workers had been killed since 1965; and three other rigs had followed the *Sea Gem*, though fortunately not with the same disastrous consequences in terms of loss of life (Carson, 1981: 150). As one member of Parliament put it, "The Bill should have been on the Statute Book a long time ago."[18] In view of the time taken over putting it there, an unusually large measure of the smugness that distinguishes so many official pronouncements on Britain's record in the field of safety legislation must surely have been required for the new Under-Secretary of State for Trade and Industry to put the bill forward in historically creditable terms:

> The whole House is excited by the prospects of the developing industries in oceanology, and we must therefore be ready to allow all these new developments to take place, but to keep alive what has been a typically British tradition—that we are ahead with our safety regulations, not in any sense to cramp expertise and initiative, but in order to ensure that those who take these initiatives are adequately protected in so far as Parliament can protect them by safety legislation. This tradition goes right back to the nineteenth century, and the House should be justly jealous to keep it up to date.[19]

Historical accolades apart, this statement conveys a fairly accurate picture of the atmosphere in which the Mineral Workings (Offshore Installations) Bill was discussed during the first half of 1971. Parliament must now legislate, not least because the inadequacies of the present system were apparent; but in doing so, it must pay due attention to the industry's technological uniqueness and progress. As Earl Ferrers explained in the Lords, "This is a new and developing industry, and the Government have been particularly anxious to adopt a flexible attitude towards the problems of it."[20] Nicholas Ridley, Under-Secretary of State at the Department of Trade and Industry, which had now absorbed the old Ministry of Power, took a very similar line in the Commons. Any archaic suggestions that regulations should be enshrined in the act itself rather than being promulgated under enabling provisions was not to be entertained "because, with this rapidly changing industry, and its advancing techniques and technology, it will be necessary to bring in new regulations and to change them frequently in response to development."[21] Reasonable enough in its own terms, this argument did not completely mollify those who seem to have been sensing here that the proposed legislation was going too far down the road toward reducing the regulation of safety to a reactive role in relation to technological advances. Such fears were expressed particularly strongly, for example, over the extensive powers of exemption envisaged in the bill, powers that some members felt might be making safety a hostage to technological fortune. On this, as on similar misgivings, however, the answer was the same: there must be flexibility "because we do not know what is going to happen in the future."[22]

Since the Mineral Workings (Offshore Installations) Act of 1971[23] was largely an enabling statute, that is, it gave the Secretary of State powers to formulate regulations covering specified aspects of offshore operations, the story of regulatory development, technological deference, and rapid progress toward self-sufficiency taking precedence over safety does not end there. While this is not the place for a detailed chronicle of the numerous sets of regulations that subsequently emerged, two points may be made. In the first place, and although there was no let up in the scramble to get Britain's oil ashore, the business of promulgating the necessary regulations proceeded at a much more leisurely or even laggardly pace. No less important, the process whereby successive sets of regulations were formulated continued to reflect a posture of technological deference on the part of an officialdom that had become heavily dependent on the offshore oil industry for rapid exploitation of Britain's offshore mineral resources.

In the first of these contexts, the gap was dramatically highlighted in early 1974 when the capsize of the *Transocean 3*—an accident that some experts claim should have taught lessons that might have averted the full horrors of the *Alexander Kielland* and *Ocean Ranger* disasters (*Glasgow Herald*, 18 February 1982)—provoked a further parliamentary debate in the British Commons. Pressed on the question of how his department, now yet again hived off into a separate Department of Energy, was coming along with the business of making regulations, Mr. Peter Emery was able to claim, justly, that four sets had already been made and, no less fairly, that two others were in an advanced state. Unfortunately, however, the regulations in operation at this stage were largely concerned with the administrative infrastructure rather than with the more detailed substance of safety requirements, and there was therefore no dodging the damaging admission that "the present method of enforcing safety requirements is by non-statutory conditions attached to petroleum exploration and production licenses granted by the Department."[24] In other words, and although he did not put it quite like this, six and a half years after the replacement of the license-linked system had been recommended, it was still in use as the main instrument of control over offshore safety. Even the two and a half years that had now elapsed since the requisite powers had been established struck some as "far too long a time to wait," frontiers of technology and understandable problems of engineering design notwithstanding.[25]

At this time, there were certainly some signs that a more jaundiced reaction was beginning to emerge with regard to the regulatory record of successive governments. Accusations of profit being put before safety, of possible corner-cutting, of a wartime attitude that treated men and materials as expendable, and of a general lack of planning being allowed to carry over into the realm of safety, were all advanced during this brief debate. Nor were they effectively refuted by countercharges of ignorance, paeans of praise for the effectiveness of an already discredited system, and bland assurances that the new Department of Energy was "completely and utterly sold on the need for adequate safety provisions and regulations."[26] Moreover, some critics were now beginning to sense something of the connection between the political economy of the North Sea and the adequacy of the legal regime governing safety. Margo MacDonald (a leading Scottish Nationalist Member), for example, linked the whole question of safety regulations to the forces underpinning rapid development and to the particular kind of government/industry relationship they had spawned:

> I would suggest that the lack of safety regulations is symptomatic of the attitude of panic by the Government towards the oil industry generally. Since I came to this House all I have heard about has been the serious economic situation. All I have heard is that the way out is for the Government to get their hands on the oil as quickly as possible, to take the oil and the revenues. That is why the Government are prepared to put exploitation of the oil before safety. The Department of Trade and Industry and now the Department of Energy appear to be permitting proper considerations of safety for those directly and indirectly involved to be bypassed in the haste to get hold of the oil and the revenues. . . . We all know that the Government are afraid to deal more firmly with the oil companies because they have the mistaken idea that the companies may pull out of the North Sea. . . . The longer the oil remains under the sea, the more valuable it becomes. We have plenty of time to make sure that we are geared up technologically, socially and physically to take advantage of the industry and make it work for us.[27]

Following the advent of a new Labour Government in February 1974, the process of imposing safety regulations on offshore oil and gas operations continued to drag on, in sharp contast with the pace of progress toward turning the North Sea into a major petroleum-producing area. Thus, for example, the date ultimately set by law for North Sea installations having to be certified as fit for their designated function turned out to be several months after the first of Britain's oil had actually started to come ashore in June 1975 (Statutory Instruments, 1974: 289). The fact that production, with its long lead-in period in terms of exploration and development, should have started before such a basic safety requirement was in operation possibly speaks for itself. Similarly, it was not until late 1975 that statutory powers were assumed for the making of regulations to cover the safety of pipelines and pipeline works, thereby plugging a gap which the Secretary of State—though not, surely, anyone familiar with the previous history of safety legislation in this field—found "astonishing."[28] Even more historically reprehensible is the fact that regulations pertaining to the provision of fire-fighting equipment were not made until 1978, with certification requirements to become operative in April 1980 (Statutory Instruments, 1978: 611). While all the technological and other difficulties must be fully acknowledged, a salutory reminder of how development was allowed to outrun safety regulation in this field is provided by the coincidence that only in the fourth month of the year in which Britian became self-sufficient in oil did it become an offense to allow anyone on to an offshore installation unless the fire-fighting equipment and plans had been officially examined and certified within the preceding two years.

Just as the process of making regulations was a slow one, so too it continued to reflect official deference toward the industry's technological status. As we have seen, such a stance had already been heralded in the debates surrounding passage to the 1971 act, while the need for detailed control through regulations rather than legislation itself had been justified on the grounds that such a rapidly changing technology would necessitate frequent changes in response to development. When it came to the making of regulations, however, the almost sacrosanct quality imputed to the industry's technology again appears to have carried the day. Reflecting the broader problems of imposing detailed requirements in technologically volatile situations, the regulations were mostly written in general terms, with detailed back-up to be provided in Guidance Notes. Once again, the rationale was rapid change, while interestingly the earlier case for a system that would allow regulations to be changed frequently in response to development seems to have been forgotten. As the Petroleum Engineering Directorate (PED) of the Department of Energy subsequently put it, "The fact that the regulations are written in general terms in a broad-brush format obviates *the need* for constant changes to keep pace with technological innovations" (emphasis added). The system of adding Guidance Notes, it insisted, "works very well for a fast-moving industry like the international oil industry" (Burgoyne Report, 1980: 228).

Much of this is not, of course, to be gainsaid. Similar problems have confronted onshore safety authorities in relation to the use of regulations, and even codes in practice, in connection with requirements likely to be subject to sector variation or technological change, while there is no doubt that Guidance Notes do provide a much more flexible mode of response in a rapidly changing situation. As far as the law relating to the North Sea was concerned, however, one immediate consequence was that the regulatory relationship became one in which attempts to impose detailed legal control over safety matters often became highly contentious. Thus, for example, when regulations were on occasion drafted with an eye to the specific, they attracted the disapprobation of the industry, as even in the case of detailed specifications for helicopter accident equipment, medical supplies to be maintained in sick bays and fire-fighting equipment for helidecks. As Esso complained, such detailed requirements were inconsistent with the more general tenor of the requirements in the remainder of the relevant regulations and might therefore be more appropriately put into Guidance Notes (Burgoyne Report, 1980: 163). But when the authorities did try in other instances to gain detailed control through such Notes, they found the industry still ready to deploy the same technological arguments and able to remind them that guidance is not quite the same thing as law. Thus, Shell later told the Burgoyne Committee of its objections to what was seen as "an attempt to place the Notes on a quasi-legal footing," to Guidance Notes which enjoined that certain sections "must be honoured," and to letters telling operators that they should formally seek relaxation from the "requirements of the Guidance Notes." Detailed Notes of this kind, the company claimed, "can have a restrictive effect on technological development offshore, particularly when they are interpreted as law" (ibid.: 132ff.). The preferred alternative was clear:

> In our view the principle that must be maintained at all times and in all respects is that Guidance Notes clarify intentions of legislation, but do not bind an operator to any particular course of action. It must be accepted by regulatory bodies that operators may elect to comply with legislative requirements by methods other than given in Guidance Notes. In such instances, it is incumbent upon the operator to justify his chosen compliance route, but equally the regulatory body must give full and fair consideration to it. (ibid.: 134)

In some ways, of course, a statement such as the above is scarcely novel. Those in the know typically resent being told how to run their affairs, while the tradition of denying the relevance and appropriateness of detailed legal controls over industrial practice is a time-honored one going right back to early factory owners, who were once described as "men of a warm temperament and a proud spirit who wish to have their own way of doing good, and who kick against any attempt to force them to do good in any other way" (Senior, 1837: 32). In the present instance, however, the germane point is that within a relationship already characterized by technological diffidence, the detailed application of legal rules had a tendency to become technologically negotiable. Nor is there much doubt that in these circumstances the companies often had the whip hand. In the case of one major North Sea platform, for example, one of the largest multinationals was apparently able to challenge both the drag coefficient specified by the Certifying Authority for the calculation of environmental loads and the "excessive" criteria laid down by Guidance Notes in relation to wind-induced tides. While no one suggested (or is suggesting) that in this particular instance the result was an unsafe structure, some people see the general context in which such negotiations take place as rather less than satisfactory. According to one chartered engineer who had been involved in the design of several platforms, the situation left quite a lot to be desired:

> Statutory Instruments (SI: the regulations) and Acts covering the North Sea exploration and exploitation leave too many areas for interpretation. Guidance Notes are not required by law and are only for guidance. The Certifying Authorities interpret the rules differently and require different standards for the same situation. . . . The outcome is that the clients can demand that par-

ticular Guidance Notes and Codes of Practice be disregarded because they are not law, resulting in the platform being designed to their requirements of cost and safety. In discussion with the Certifying Authority . . . they agree that only SI shall be used as a guidance. A particular instance recently resulted in the client demanding that the fire protection/fire detection be operated manually only, for one of the largest proposed platforms in the North Sea. Fire codes, Department of Energy Guidance Notes, and all other guidance and codes of practice were not permitted.

If negotiations on this one-to-one level represent one setting in which the industry's technology facilitated the growth of a special regulatory relationship, bargaining over rules and regulations on a more industry-wide basis was another. Thus, for example, in 1974 the Under-Secretary of State for Energy explained that the consultations that had taken place with the industry on the regulations drafted up to that point had been "as great as with any legislation I have known, mainly because we are dealing with evolving technology and because, in construction and design, we are working with conditions never before faced when using rigs or building platforms."[29] Closer to the present, the controversial Guidance Notes on fire-fighting equipment are said to have gone backwards and forwards ten times before finally being accepted, one of the changes being the substitution of "should be honoured" for "must be honoured." According to one company safety officer, they were "probably the most fully discussed Guidance Notes ever issued," while he was at pains to anticipate obvious criticisms by an appeal to the now familiar canons of technological progress: "While there may be criticism from some about their lack of definite stated rigid parameters and type of equipment, we have always felt that flexibility is the key note, to allow for technical developments" (*Oilman*, 23 August 1980).

Finally in this context, it must be remembered that whatever the issues at stake, the negotiations surrounding offshore safety regulations took place within a relationship that involved the authorities on one side and one of the world's most powerful industries on the other. Nor did the latter comprise just an amorphous array of companies to which officials had to relate on the nebulous basis of speculations as to what the industry might collectively define as its interest. As far as North Sea operations were concerned, the companies formed themselves into a coherent and effective organization, the United Kingdom North Sea Operators' Committee, shortly after the first round of licensing in 1964.

Government welcomed the formation of this body and "increasingly found it a most useful one to consult as providing an appropriate voice for the total offshore oil industry"—this notwithstanding the fact that membership was restricted to operators and thereby excluded the subcontracting side of the industry, which then, as now, employed the great majority of North Sea workers. This exclusiveness continued after 1973, when the Committee was transformed into the fully registered United Kingdom Offshore Operators' Association, as did a feeling on the part of the British and other "independents" that the organization was unduly dominated by the large multinationals. Thus, negotiations on safety as well as other matters took place within a context where the paramount industry voice was always likely to be that of the large international companies, upon which successive governments had become so substantially dependent. "The main channel of communication with the offshore oil industry," reported PED in 1979, "is through the United Kingdom Offshore Operators' Association . . . who represent all the companies operating in the United Kingdom Continental Shelf" (Burgoyne Report II, 1980: 229). "What we do," explained one official, "is negotiate with the industry: after all, we do still live in a democracy."

What such a statement ignores is the fact that even in a democratic context, "negotiation" involves the crucial dimension of power, and that all the trappings of sovereignty notwithstanding, it is by no means clear that on this score the authorities necessarily had the upper hand. Committed to rapid development, for reasons stemming from broader considerations of political economy, successive governments had become unduly dependent on the goodwill of the international oil industry which alone could secure politically defined objectives within the preferred time-scale. In consequence, they had to tread warily in the field of safety, as in other areas such as depletion, taxation, and public participation, lest multinational capital should invoke the nineteenth-century maxim that "like love, its workings must be free as air; for at sight of human ties, it will spread light wings . . . and fly away from bondage" (Ure, 1835: 435). More prosaically, the development of statutory controls over offshore operations took place against a backcloth of technological hegemony exercised by the international oil industry, a hegemony that could only be challenged at the risk of slowing down the offshore enterprise. As a result, both legislation and regulations were framed in such a way as to avoid interference with the industry's own technological trajectory, and in this way technological issues played a crucial role as the mediating link between the political economy of speed and the dilatory emergence of an adequate safety regime.

INFORMATION DEFICITS, INSTITUTIONAL STRUCTURES, AND PRIVATIZATION

As we saw at an earlier point, one important respect in which the British authorities were over a barrel, so to speak, with regard to North Sea developments was that their commitment to speed had generated radical deficiencies in both basic information and expertise. However, just as factors rooted in the political economy of the North Sea exacerbated official difficulties in connection with general information, so too they inevitably had an effect on the amount and quality of information related to the more specific issue of safety. Nowhere was this clearer, for example, than in the context of knowledge about the implications of the operational conditions to be expected in such an environment, for here the industry itself was relatively uninformed. Initially, by all accounts, the companies simply extrapolated from their previous experience in other parts of the world, principally from the operations that had been going on for many years in the Gulf of Mexico, and, as a result, they drastically underestimated the magnitude of the problems entailed in operating on the North Sea's Continental Shelf.

While this problem may not have been particularly severe as long as operations were confined to the easier waters of the Southern Basin (which, nonetheless, did see its crop of accidents), the difficulties became acute when exploration moved northwards into the deeper and much stormier waters off the coast of Scotland. There "conditions which many Americans did not realize existed"[30] led to new problems of design, construction, and monitoring as attempts were made to scale up existing technology to meet a more hostile environment. The inadequacy of the basic information available at this stage is evident from the following extract from one report on a symposium held in connection with the proposed Construction and Survey Regulations in 1973:

There are already insufficient data for the North Sea, and theories are as many in number as there are experts to expound them. Who, then, can be expected to say with certainty which one is correct . . . ? Mr. Campbell (Shell) reasserted the "frontier technology" situation of the North Sea and emphasized the lack of data, particularly in predictions of wave height, wave forces, current and therefore stresses and fatigue life of structures. (*Offshore Services*, July 1973: 17)

If the industry itself lacked information of such vital relevance to safety, the authorities were no better off. As Noreng remarks, perhaps generously, the lack of systematic data at this point has to be set against a background of neglect in research and development effort that must be laid mainly at the door of governments which, "until the cost escalation reached fairly alarming proportions in 1975–76 . . . seemed to have had an uncritical faith in the ability of the oil industry to master the problems of development." More than that, he argues, in the case of British governments the balance-of-payments predicament provided good reason for pushing development and for allocating high priority to time targets, with the result (he implies) that there was little incentive to interfere with the companies' practice of foreshortening and even telescoping different stages such as planning, design, construction, and installation (Noreng, 1980: 95ff.). The implications for safety are fairly apparent. As one eminent metallurgist later remarked, albeit in the aftermath of the *Alexander Kielland* disaster, "It would be much more sensible if research was done ahead of design, but in the case of oil I believe we are in a great hurry because of the energy crisis and that often the rules are worked out on the sparsest of information" (*Oilman*, 5 April 1980).

If such comments must necessarily smack of wisdom after the event, there were some people who were skeptical enough at the time. According to one critic in 1974, developments like the Brent and Forties fields were a bit like putting the Concorde in flight without a flight-test program, and while he may have been somewhat reassured to learn that the government was currently spending around 1.5 million pounds sterling on research and development, his misgivings about the authorities' role in generating the information requisite to the safety of such undertakings were not entirely allayed:

It is true that the operators and licensees will have placed upon them, individually and collectively, an obligation continuously to monitor the competence of these structures in operation. There does not appear to me to be any desire on the part of the Government to sponsor research which would seek to anticipate such behavior. It is essential that we try to get Government-sponsored research and development, in association with the industry, to anticipate the conditions under which these structures are likely to operate.[31]

Such calls for government-industry cooperation in the acquisition of vital safety information might have been couched in somewhat less optimistic terms if they had taken account of industry's remarkable penchant for secrecy. To many companies, one of the great incidental advantages of the North Sea was the opportunity it offered for the development of an offshore technology that would stand them in good stead when dwindling reserves might increasingly drive them to more difficult maritime locations. Not surprisingly, they were therefore less than keen to share their growing knowledge with anyone. Thus, for example, it was claimed in 1973 that other firms engaged in "frontier-technology" developments were pleading with the oil companies to release up-to-the-minute data, but to no avail, while a representative of the Institute of Oceanographic Sciences "decried oil companies' policy of keeping secret their environmental data." Significantly, he was also reported to have "deplored the fact that platforms are being designed using data which is not available for public scrutiny and discussion" (*Offshore Services*, July 1973: 17ff.).

All this, of course, was some time ago. However, despite the distinct improvements that took place after 1974, there is still reason to believe that an improved flow of information requisite to safety was not all that easily accomplished. Such difficulties were particularly evident, for example, in the case of failures, where information is obviously vital both for purposes of research and for the dissemination of cautionary data. The incentive to maintain production, it has been said, meant that failures might on occasion be quickly rectified but not investigated, with the consequence that the necessary information was not there to be communicated in the first place (Burgoyne Report, 1980: 115). Moreover, such events often went unreported to the Department of Energy, particularly when no injury was involved. Indeed, according to one Certifying Authority, they were not reported "in a general sense unless they [led] to or [were] part of accidents which cannot escape public detection, e.g., a blow-out or the sinking of a platform." One result, claimed the same organization, was that institutions engaged in the vital business of fatigue research "cannot test schemes for crack propagation estimates for lack of examples on which to apply them." If this is a highly specific instance of informational deficit on the research plane, a more general difficulty became apparent when, in 1978–79, the Burgoyne Committee was unsuccessful in obtaining information about the extent of expenditure by the industry on offshore safety research. As a result, it was left with little more than pious hopes and exhortation:

> The offshore oil industry of the United Kingdom has the opportunity to lead the world in technologies developed to meet the particular challenge of the North Sea. No doubt it foresees benefits from the transfer of such technologies to other countries. We hope the industry will agree with us that this justified not only a considerable effort of research into the general technologies of exploration and production, but also a proportionate and related research effort devoted to engineering and occupational safety. (Burgoyne Report, 1980: 49–50)

Against this background of rapid development, sparse information, and industry cards played close to the chest, the authorities had to come very much from behind in terms of marshalling their own expertise with regard to offshore safety. And, inevitably, the process was a slow one. Thus, for example, the first two inspectors were not appointed until 1966 and 1968, respectively, while an identifiable "inspectorate" did not make its appearance until 1969. By the time the Mineral Workings (Offshore Installations) Act was passed in 1971, the staff involved in safety and operations, including pipelines, still only totaled three. As late as January 1973, the Department of Trade and Industry was still confessing to some worries about the rate at which it had built up the "technical resources [that] are not required just for the cases we are talking about here, which is mainly assessment of prospects and the evaluation of fields and helping in forming policy, but also for other activities like regulation, safety, inspection and so forth" (Committee of Public Accounts, 1973: 130). By that point, the total staff had risen to seven, but there were still significant gaps, not least in relation to diving, which did not attract its first appointment until 1974.

The same year saw another development of considerable significance both for the informational problem and for the offshore safety regime as a whole, the establishment of a separate Department of Energy. Representing what Noreng calls a vertical administrative structure as opposed to "a horizontal structure consisting of hierarchical layers of Government, agencies with generalized functions" (already existing onshore agencies which would simply extend their domain of responsibility sideways to incorporate relevant aspects of offshore operations), this system offered what the same author sees as distinct advantages. For one thing, it increased the "detailed insight" available to a government which, he agrees, "had inferior information resources compared to the industry" (Noreng, 1980: 136). More generally its primacy in terms of controlling oil policy gave decision-makers the power to pursue policies of optimal efficiency over a functionally specific field. To Noreng, recourse to vertical structuring is almost the *sine qua non* of a rational policy with regard to North Sea oil, whether British or Norwegian:

> In both countries the main responsibility and powers are embodied in the vertically organized primary structures. The powers give the organizations of the primary structure the freedom of action required to make rational decisions, weigh preferences and have the necessary insight and knowledge to maximize expected utility. This is in many ways also a prerequisite for the development and implementation of a rational oil policy. (Noreng, 1980: 141–42)

Despite its undeniable advantages, however, this vertical structure also involved some important drawbacks. At a general level, for example, it meant that responsibility for offshore safety was placed in the hands of a Department that not only had a parallel responsibility for production but also had the objective of rapid development of the North Sea as part of its very *raison d'être* (see *The Times*, 9 January 1974). At a time when the oversight of safety in nearly all onshore industries was being taken away from "sponsoring" departments and was being entrusted to a horizontally structured or generic regulatory agency (the Health and Safety Commission and its Executive),[32] this combination of functions within one component of the administrative machine could only exacerbate the risk that the regulatory system would be contaminated by the increasingly exigent considerations emanating from the political economy of the North Sea. Whereas in Norway the trend was somewhat different, because of fears that "a control function kept within the Ministry [of Oil and Energy] might have been more exposed to political pressures" (Noreng, 1980: 141), in Britain responsibility for offshore safety was located at exactly the point where such pressures were likely to be most acute. Nor should we forget that the thrust of such pressures right up through the 1970s was toward the fulfillment of short-term objectives through rapid exploitation, the longer term considerations uneasily embraced by Labour from 1974 onwards notwithstanding. Appropriately enough, it fell to a former Minister in that government to voice the objections in principle to such a structure, albeit after inconclusive, if creditable, efforts to break the Department of Energy's monopoly had been made:

> What we are saying . . . is that a fundamental principle is being breached in an industry that is of great concern to the House because the Department of Energy is to be solely responsible for health and safety. It has many other responsibilities for the offshore industry—to get out as much oil as possible, to get the revenues and, under pressure from the Treasury, to consider the interests of the oil companies. We categorically and emphatically say that making health and safety its sole responsibility is insufficient.[33]

This potential for conflicting priorities within vertically organized administrative structures does not receive much attention in the Noreng analysis referred to above, although its author does concede that more competition between generalized agencies and the primary structures would probably benefit safety and working conditions. On other fronts, however, the same author is at pains to stress that the advantages to be derived from vertical structuring have always to be offset against fairly evident drawbacks. Thus, for example, he recognizes that gains in micro-effectiveness may well entail deficits at the macro level, while the possibility that specialized agencies may develop their own culture and depart from generally established procedures is fully acknowledged. So too is the risk that practices in interpreting and enforcing rules may become institutionally idiosyncratic. Most of all, perhaps, Noreng emphasizes that, although vertical structuring does indeed enhance detailed control and access to information, it also means very close contact between controllers and controlled. As a result, it brings in its train the attendant risks of the agency's being "colonized," of its coming to act as an advocate for those whom it supervises and, indeed, even of control being reversed (Noreng, 1980: 113ff.).

With recognition of possibilities such as these, we are brought back to fairly well-charted sociological waters. The tension between micro and macro levels, for example, is a well-rehearsed topic, even if the emphasis is rather more on the possible contradictions between *ad hoc* bureaucratic decision-making and overall planning (see Scott, 1979:

156ff.). Similarly, the capacity of monopolies to "occupy" specific parts of the diffuse structure comprising the machinery of the state has often been commented on, particularly in connection with science and technology (see Hirsch, 1978: 101). If this is the analogue of Noreng's extreme category of "colonization," the more general tendency has been cogently described by J. Habermas:

> It is possible to show that the authorities, with little informational and planning capacity and insufficient co-ordination among themselves, are dependent on the flow of information from their clients. They are thus unable to preserve the distance from them necessary for indepedent decisions. Individual sectors of the economy can, as it were, privatize parts of the public administration, thus displacing the competition between individual social interests into the state apparatus. (1976: 62)

Elsewhere I have argued that just such a process of displacement came to play an important part in making the safety of North Sea operations the subject of considerable conflict between different agencies of the British state during the latter part of the 1970s. Here, however, the point to be emphasized is that the way in which the authorities made up lost ground in terms of safety expertise did indeed invoke the generation of a particularly close relationship between those whose business was control and the industry to be controlled. While the precise nature and implications of this relationship subsequently became the subject of bitter dispute, even then protagonists on all sides would seem to be *ad idem* on the appropriateness of a family metaphor. Thus, in 1979 the director of the Department of Energy's Petroleum Engineering Directorate could look back almost with nostalgia to the days of youthful discovery, when "the industry was in its infancy and both it and the then Inspectorate were learning." Switching family roles to that of the parent with a wayward child, he went on to describe the current situation as one in which "the industry is in its lusty youth and may now need the sharp corrective of an occasional prosecution." As for the effect of those formative years when the rules were still in the making, the diffidence of youth had clearly not been entirely forgotten:

> In the early days they did not have to administer regulations, they had to create them. Because many PED inspectors were involved in the writing of the regulations, there is no doubt that they regard regulations in a different light

from the Factory Inspectorate—not as infallible laws which must be obeyed, more as the product of fallible human beings—themselves.

Resort to the metaphor of family was not, however, restricted solely to officials. Shell, for example, invoked a similar image by recalling that "the Department of Energy Inspectorate has grown up with the offshore industry and has acquired valuable knowledge and experience" (Burgoyne Report, 1980: 137), while in others such closeness provoked "fear that this intimate relationship that has developed will inhibit it (the Department) from making criticisms and from imposing costs and other matters."[34] Nor were such fears assuaged by the fact that there was considerable traffic in personnel between the regulatory agency and the regulated. In one direction, the authorities in many cases acquired their expertise from the most obvious source, inspectors being recruited, as PED later admitted, "for their experience in either the oil or a related industry"; in the other, and although the Department is obviously coy about publishing the statistics, there was also "wastage" from the Department of Energy back to the industry, a tendency that was epitomized *in extremis* when the Head of Operations and Safety resigned to become permanent Technical Secretary to the United Kingdom Offshore Operators' Association. From such intercourse would arise charges of an even less wholesome family relationship: "The Minister should understand that some of us feel deeply that an improper relationship is involved. Indeed, it is an incestuous relationship. The sponsoring Department has no business getting involved with health and safety."[35]

Although it is obviously impossible to be concrete about the effect which such close relationships, improper or otherwise, may have had on the rule-making process, there is little doubt that they did bolster the industry's confidence in the regulatory stance adopted by the PED. As the general manager of one "independent" operator explained, the British system was superior to the Norwegian in this respect, since Norway, "a totally socialist state," had destroyed individual enterprise and had fallen back on looking after its workforce "from cradle to grave" by means of "silly little bureaucratic regulations." In contrast, the British were described as much more sensible because "the men at the top have experience of the industry and know the problems." Some way down from the top, PED inspectors themselves seem in the past to have placed considerable value on such experience. According to one of them, there was even a feeling that unless you had worked in the industry you could not purport to know anything about it, much less to undertake research in the area. As for the industry itself, little secret is made of its view that inspectors should ideally be drawn from its ranks. To Shell, an increase in the Department's qualified staff "with a suitable background of experience in the offshore oil industry" would help to strengthen "existing links" and to "facilitate the development of new codes and standards" (Burgoyne Report, 1980: 138). Not surprisingly, perhaps, others take a rather different view:

> Unions have pointed out that while it can be argued that the only significant expertise in offshore safety lies with either the Department of Energy or the Companies, the interchange of personnel between the two may raise certain issues. For example, while it is certainly not intended to question the integrity of any individual involved, the possibility of shared values and closed groups amongst offshore personnel may be used as an argument to question the independence of inspectors who have a close association with the industry. (ibid.: 292–93)

Close relationships, even if described in the more enigmatic terms of "shared values and closed groups" rather than outright impropriety, did then raise doubts about the extent to which the regulatory agency in question was able to maintain its distance from the industry it had responsibility for controlling. While the speed of North Sea developments posed a constant problem in terms of the adequacy of official information and expertise, the machinery and means whereby the latter were acquired placed offshore safety inside something of a regulatory ghetto. "Our companies" was the constantly repeated phrase used by one senior Department of Energy official when referring to the offshore oil industry in interview, while Tony Benn later admitted that the tendency for "the thing to get a bit cosy" had been particularly noticeable during his stay at the Department. More generally at many points in this research instances were encountered where personnel from the PED did indeed seem to be acting more as the industry's advocates than as its overseers. In one particularly pointed case, when members of a commercially organized seminar were told that the offshore safety record was not a good one, it was not the oil men present but the Department of Energy representative who rose to refute this allegation against an industry "that has come further in seven years than any other comparable one has in three times that time." Appropriately enough, suggestions that the Department's own approach might be something less than stringent provoked an industry participant to reply that the speaker's remarks were "less than generous" not only to the industry but also to the Department of Energy.

Close or even closed associations were not the only factor underpinning the growth of a special regulatory relationship during the period with which this paper is mainly concerned, however. As we saw above, a recurrent theme in debates about offshore safety had been the need to avoid any steps that might hamper the industry's technological progress, upon which depended the continued momentum of North Sea development, particularly as the enterprise moved into more and more difficult terrain. Thus, the commitment to speed elevated the industry's technology to almost sacrosanct status, and, not surprisingly, this too was to have a pronounced effect on the regulatory relationship which became established. In this respect, as in so much else, the authorities found that safety controls were indeed hostage to technological fortune in more senses than one.

At the most obvious level, the way in which the industry was permitted, and even encouraged, to press ahead with operations right up to technological limits where so much was unknown, although not unknowable, meant that safety administration understandably became preoccupied with the prevention of outright catastrophe. As the head of PED explained, ever since the loss of the *Sea Gem* in 1965, the thinking of relevant government departments had been dominated by concern to prevent a major disaster as a result of structural failure, blow-out, fire, or explosion, with the possible consequence that "insufficient attention" was being paid to "prevention of accidents of an occupational type." Thus—and while it must be underlined that there is nothing in the British sector's record so far to detract from any claims that might be made for the success of this strategy—the more mundane type of accident which, as suggested earlier, accounts for the bulk of North Sea casualties, was arguably accorded less priority than its incidence merited. Not only that: the "disaster orientation" of the authorities therefore also came to parallel the industry's own approach, which was always primarily concerned with the more economically and politically consequential possibility of a catastrophe taking place on its own technological boundaries. In short, the image of offshore danger as an inevitable consequence of technological trail blazing was reinforced by the institutional responses to rapid development.

Such preoccupations obviously did little to expose the industry to the healthy skepticism which would insist that on top of the undoubtedly special risks associated with speedy North Sea development, attention should also be focused on a wide range of concerns in which offshore risks are not unique. Nor was the administrative structure already described very likely to have this effect. Perpetuated by concern to maintain rapid progress, this specialized structure served to confirm the industry's sense of uniqueness by isolating the regulation of offshore safety from developments in other sectors of industry. Thus, the input of official experience, in terms of lessons learned from the history of controlling other industrial activities like construction or even manufacturing, seems to have been relatively limited, and the industry was able to continue in the belief that the safety problems about which it had to worry were, above all else, special. Even cross-fertilization from other technologically sophisticated industries seems to have been fairly restricted; the experiences of the nuclear, aerospace, and chemical industries of risk and safety assessment were particularly under-utilized. More generally one of the Certifying Authorities in 1979 noted a reluctance to learn from others:

> Since the *Flixborough* and other disasters, the interest in safety technology has heightened and more industrial organizations are becoming conscious of the benefits to be gained from its application. Regrettably, the oil industry has not yet generally recognized that the experience gained in other industries in safety engineering can equally well be applied to the offshore oil industry. (Burgoyne Report: 154)

CONCLUSION

In this paper I have attempted to show how issues involving technology, technical expertise, and the like played an important role in shaping the legal response to the occupational dangers encountered in the course of developing Britain's offshore oilfields. Starting from a counsel of caution to the effect that the imagery of technological frontiers should not precipitate premature legal capitulation, particularly since the majority of North Sea accidents to date have emanated from relatively mundane and regulatable factors, I went on to locate discussion of the role played by technology with a broader, if necessarily cursory, examination of the political economy. Against this background, it was suggested that the high priority allocated to speedy development of the Continental Shelf's resources both pushed the formulation of statutory controls into a rather poor second place and, no less importantly, hedged the regulatory process around with a series of constraints stemming from the desire to avoid obstruction of the industry's technological progress. Similarly the same concatenation of forces produced an inevitable information deficit on the part of the relevant authorities, a gap that played no small part in the emergence of a specialized administrative structure for handling offshore affairs, including safety, and in the growth of a privatized relationship between controllers and controlled. Although it cannot be elaborated on here, the outcome was a regulatory regime which, in both principle and practice, fell far short of an adequate legal response to the hazards of offshore employment. In principle, the fusion of functions within one Department produced an unhealthy measure of administrative propinquity between issues involving safety and those concerning production. In practice, the consequences of such an arrangement can be seen all too clearly in the following extract from a report prepared by an inspector after a visit to one of the North Sea's largest production platforms:

> The economic plight of the country and the political significance of producing oil yesterday are understood and appreciated, but in my opinion, from a purely safety aspect "start-up" has been three months too soon. I state this so that no one is in any doubt that corners are being cut and calculated risks are being taken to obtain objectives!

Finally, it is appropriate to ask whether any lessons can be learned from this somewhat sorry tale of attempts to grapple with the offshore oil industry, its technology and its power. For Britain itself, not very many perhaps, although there is still a pressing need for a radical restructuring of safety administration even this late in the day, and there still remain a number of residual problems about offshore law enforcement which should and could be sorted out. As far as other countries are concerned, however, there is for once an opportunity to learn something from British mistakes rather than from its more customary posturing as self-evidently superior. In particular, for countries such as Canada and Australia, which now find themselves at a stage in offshore development somewhat similar to that of Britain in the 1960s and early 1970s, there are valuable insights to be gained from the catalogue of British failure.

It is my hope that this paper has gone some way toward setting out just what such insights may be.

NOTES

1. This story is recounted by Dr. J. Kitchen (1977) who collaborated with the author during a preliminary investigation to assess the feasibility of research in this area.

2. These figures are based on the statistics published annually by the Department of Energy in *Development of the Oil and Gas Resources of the United Kingdom* (referred to as "The Brown Book"), London, HMSO, 1974.

3. In August 1981, two helicopters engaged in North Sea operations crashed on successive days, killing one and thirteen people, respectively. Thus, the pattern of improvement over the decade is broken in the 1981 statistics.

4. Apart from one of the helicopter accidents referred to in note 3, the only other multi-death accident to have occurred in the British sector of the North Sea was the loss of the drilling rig *Sea Gem* in 1965, with the loss of thirteen lives.

5. But see *Offshore Safety* (the Burgoyne Committee Report), Cmmd. 8941, London, HMSO, 1980, p. 114.

6. Neither the Norwegian text nor the English summary makes it absolutely clear whether design and fabrication were completely satisfactory in this respect. On balance, however, the most plausible interpretation is that failure to close the doors was the issue at stake.

7. Personal communication.

8. See, for example, P. Nore, "International oil in Norway," in J. Faundez and S. Piccioto, *The Nationalisation of Multinationals in Peripheral Economies*, London, Macmillan, 1978, p. 168.

9. *Parliamentary Debates* (UK) (Commons), Volume 991, 1980–81, 6 November 1980, col. 1478.

10. *Parliamentary Debates* (Commons: Standing Committee A), Volume 1, 1963–64, 19 February 1964, col. 78.

11. *Parliamentary Debates* (Commons), Volume 688, 1963–64, 28 January 1964, col. 224.

12. *Parliamentary Debates* (Commons: Standing Committee A), Volume 1, 1963–64, 9 February 1964, col. 76.

13. Ibid., 4 March 1964, col. 136.

14. *Parliamentary Debates* (Commons), Volume 642, 1963–64, 7 April 1964, cols. 876–77.

15. Ibid., col. 29, Sec. 3.

16. Alternatively, the Minister could effect entry and cause any necessary rectifications to be carried out at the licencees' expense.

17. *Parliamentary Debates* (Lords), Volume 254, 1963–64, 19 December 1963, col. 394.

18. *Parliamentary Debates* (Commons), Volume 816, 1970–71, 28 April 1971, col. 650.

19. Ibid., col. 649.

20. *Parliamentary Debates* (Lords), Volume 315, 1971, 18 February 1971, col. 743.

21. *Parliamentary Debates* (Commons), Volume 816, 1970–71, 28 April 1971, col. 648.

22. *Parliamentary Debates* (Commons: Standing Committee G), Volume 4, 1970–71, 17 June 1971, cols. 15–16.

23. Ibid., 1971, col. 61.

24. *Parliamentary Debates* (Commons), Volume 867, 1973–74, 16 January 1974, cols. 690–91

25. Ibid., col. 679.

26. Ibid., col. 690.

27. Ibid., col. 681–82.

28. *Parliamentary Debates* (Commons), Volume 891, 1974–75, 30 April 1974, col. 493.

29. *Parliamentary Debates* (Commons), Volume 867, 1973–74, 16 January 1974, col. 691.

30. *Parliamentary Debates* (Commons), Volume 867, 1973–74, 16 January 1974, col. 691.

31. *Parliamentary Debates* (Commons), Volume 867, 1973–74, 16 January 1974, col. 693.

32. Under the auspices of the *Health and Safety at Work, etc. Act*, 1974, col. 37.

33. *Parliamentary Debates* (Commons), Volume 991, 1980–81, 6 November 1980, col. 1493.

34. *Parliamentary Debates* (Commons), Volume 991, 1980–81, 6 November 1980, col. 1643.

35. Ibid., col. 1515.

7

Counterpoint: Oil Development in Britain and Canada

LARRY PRATT

The story of how Britain developed the North Sea oil in the 1970s has been told by a number of social critics. Professor Carson's work admirably describes how the pace of development in Britain contrasted that of Norway. Carson relates that Britain had an exaggerated view of the financial and economic benefits of North Sea oil. This arose because there was little on the economic horizon that looked good for Britain. Oil was the only possible financial salvation; the benefits were grossly exaggerated. At the same time, the costs in human lives of developing petroleum resources and the problems that would arise from bad timing were underestimated. Other literature emphasized such costs as lost economic rent and the bargain prices on early British oil. In addition, Britain lost many of the spin-off industrial benefits by again encouraging an overly rapid pace of development. But Carson's focus is the cost of oil measured in needless deaths and injuries incurred by the speed of development. Although the Norwegians have clearly made mistakes and have suffered serious accidents, they did have a more realistic view of oil; they saw it as a curse as well as a blessing. Given Norway's small, traditional type of society, approval for development was advanced under a regime of careful regulation and control; the timing of development and the rate of depletion were carefully monitored. The strength of Carson's investigation lies in its ability to link empirical observations with the broad theoretical issues. Carson follows the linkage developed by C. Wright Mills between personal troubles and "the big ups and downs," the broad transformations of society.

This is a model for his study. This attempt to link the problems of safety on the rigs with the energy crisis, the development of OPEC (Organization of Petroleum Exporting Countries), the crisis of the world capitalist economy, and Britain's specific economic problems in the 1960s and 1970s succeeds well. Carson explains the course of British petroleum development in the context of combined and uneven development of international capital and geopolitical changes in energy sources. This is a difficult analysis to undertake, but Carson succeeds admirably.

The second strength of Carson's paper lies in its emphasis on class analysis. Most studies of oil, including my own work, completely ignore the role of labor, the working class, in the development of resources. The reason for this lack are not difficult to discover. Petroleum development is a very capital- as opposed to labor-intensive resource. In many jurisdictions, including the North Sea, the working force is not organized. The unions have been excluded from many sectors of the oil industry. Critical analysis has tended to focus on capital and on rent and the creation of a rentier type of society. The state which owns the resource lands in the offshore concentrates wealth by siphoning a tax levy from the profits. In contrast to this type of analysis, Carson insists that one of the chief reasons for the offshore safety record has been the relative weakness of the working class in attempting to organize the rigs and the inability of organized labor to mobilize the bureaucracy which, onshore, would ordinarily control safety.

Finally, for me as a political scientist the research is important because of its sound approach to the nature of the study of the state. Carson insists on the empirical study of the British state, showing that it is a very complex set of institutions. It is neither purely autonomous, independent, or neutral; nor is it simply the pliant instrument of multinational oil companies. As Carson points out, there is an element of dependency, heightened dependency even, but there is also autonomy in certain areas. He points out that the real pressures for development in the North Sea emanated from various sectors of government itself, particularly from Treasury, which knew the potential for revenue. However, Carson fails to acknowledge that the multinational companies after the early 1970s also had a strong interest in developing alternative sources of supply outside of OPEC. By the early 1970s, the process of Arab nationalization at the source of production and the gradual takeover of the oil industry was well underway. Control in the industry was passing to governments. I think it is important to realize that the oil industry has always disciplined producing governments by creating surplus capacity and by creating new external and competing sources of supply to try to weaken the bargaining power of those who own oil resources. Consequently the multinationals had an interest in bringing on supplies in areas like the North Sea; in fact, the North Sea oil is now one of the principal competitive sources to OPEC and one of the key reasons why the power of OPEC has been substantially reduced in the past several years. In other words OPEC was disciplined in some measure by the development of the North Sea resources.

If we look at Canada, I think that Professor Carson is right to insist that we are not learning well the lessons from overseas. We are pushing ahead, pushing the development of frontier resources in areas about which we have extremely little scientific knowledge and in which there is not a good working regulatory system for coping with the environmental effects of development, or as the *Ocean Ranger* tragedy shows, the safety features. Clearly the national push for self-sufficiency in energy sources has become a promise of regional salvation, especially in the depressed Atlantic provinces. Safety questions take a back seat to the promise of jobs and profit. Politicians push for the technological benefits of development whether in the Beaufort Sea or the North Atlantic. Contracts are sought for ice-breaking tankers and other supply services. Technological imperatives take priority over safety and environmental factors. Once again the impetus for development comes from governments, governments that are pushing for economic development and that must drag a reluctant private sector into the frontiers with generous tax incentives.

I am currently working on a study of Petro Canada, the Canadian national oil company. If one were to ask someone at Petro Canada or at the Federal Department of Energy, Mines and Resources about the pace of development in frontiers, they would say that it has been much too slow. If you look at the statistics between about 1968 and 1973, there was a fairly steady increase in exploratory drilling in areas like the McKenzie Valley, the High Arctic, the Eastern Arctic Islands, and the Sable Island area, and a little bit off Newfoundland as well. One of the major impetuses for this activity was the discovery of oil at Prudhoe Bay in Alaska. This was certainly a large part of the thrust into the Northern Arctic region, but there were also political imperatives or political factors. The federal government was interested in establishing sovereignty claims in the Arctic Islands, which is one reason why it backed Pan Arctic Oils and its exploration activities in the high Arctic. After 1973 the pace of exploratory drilling dropped off very dramatically in all of these areas. This change was partly because the oil industry had not found anything significant, partly because of complications over the MacKenzie Valley Pipeline and uncertainty over whether that project would go ahead, and partly because of the conflict between Newfoundland and the Canadian government over jurisdiction and ownership of offshore resources. Also in the mid 1970s the industry experienced new discoveries in Alberta which tended to shift activity back to the conventional basins in western Canada. This is one of the principal differences between Britain and Canada.

There is also the problem of divided jurisdictions that arises in federal systems which creates investor hesitancy because of ambiguous political responsibilities and powers. In addition, Canada, unlike Britain, has had an established and conventional oil and gas industry for thirty-five years or so, which to some extent has meant that the western conventional industries have competed for investment with the frontiers. In turn, this fosters government competition to create rival climates for investment. Nevertheless, the frontier investments have been precarious at the best of times due to the unfamiliarity of offshore conditions. In the 1970s the conservative petroleum industry invested about 85 percent of its exploration firms in the West instead of the frontiers. Not surprisingly, the federal government has sought to offset this risk aversion by encouraging the industry to become more heavily involved in the frontiers through tax relief for frontier investment. Since about 1973 the federal government has been trying to get the industry interested in the High Arctic, in the Beaufort Sea, and in the Atlantic offshore. Petro Canada was created for this very reason: to stimulate the exploration and development of the frontiers. The federal government has a notion that is sometimes referred to as the "nexus or the link" between exploration and production, and this is done by creating a big state company. According to this theory, one can go out and look for oil, but if it is found, it will not necessarily be produced. In practice, Petro Canada has become just as eager for rapid development as any private developer. The reasons are not difficult to find. Petro Canada was given an extremely costly mandate to explore for oil and gas and has carried out that mandate. It has spent an enormous amount of money in the frontiers; at the same time, it has been told to become financially self-sufficient. Due to a heavy need for cash, the company wants early development of the frontiers, particularly early development of Hibernia and the Atlantic offshore. Paradoxically, the government created an institution that is designed to speed up exploration and in effect is at least as interested as the private sector in rapid development of the frontiers.

An analysis of Petro Canada's current capital budget reveals that this Crown corporation tops the industry in expenditures. It has the largest capital budget of any oil company in Canada. In 1982 it was spending $400 million in frontier exploration, about one-quarter of its budget. It seems to me, although I cannot prove it, that this goes well beyond what would be prudently required for self-sufficiency. It seems that the level of exploration, especially in the Atlantic offshore, is not really being determined by requirements for self-sufficiency but for the exploration and development of the deposits lying offshore with an eye to the American export market.

In other words a commercial interest, not a policy interest, lies behind this heavy emphasis on the frontiers. In addition, an industry competition factor has accelerated the pace of development. The prospect of exports has encouraged rivalry between Dome Petroleum and Petro Canada to see who can get into production first, whether it will be the Beaufort Sea in the Arctic or the Atlantic offshore fields. A rivalry has arisen over shipyard capacity in order to guarantee priority in production. Consequently commercial pressures on the state to hasten development are exerted partly through its own company.

The *Ocean Ranger* tragedy obviously has some serious implications not only for Petro Canada but also for its partners there. Petro Canada carries out all of its ventures with private companies and ironically shares the same kind of commercial interests. Professor Carson's work raises this paradox of state interests. Rather than finding a monolithic state structure which is led by the nose by powerful multinational corporations or, indeed, by powerful national companies, Carson draws our attention to the many, and sometimes conflicting, parts of the bureaucracy pursuing different interests. Neither of us subscribes to a simple pluralistic interpretation of the state. The picture that emerges is a very complicated one. What must be done is to find out how one gets organizational power or interest to support safety claims and environmental claims to give those the priority that is currently being given to commercial production. Clearly this takes us back into the realm of class analysis and an examination of who works on the rigs, who owns and operates them, who gets the wealth, and how much power each has over the conditions of work and production.

IV

The Media, Technology, and the Environment

The role of the media in affecting public awareness of environmental questions has already been alluded to in Professor Carson's discussion of the journalistic imagery of "frontier technology" in the offshore petroleum industry. The idea of men struggling to overcome the unknown on a dangerous technological frontier, however, diverts attention from the social factors that imperil worker safety. The role of the media is dealt with in this section by two key papers. Allan Mazur describes how the mass media operate in envirionmental controversies. The mass media are differentially "plugged into" news sources and in cases of specialized scientific and technical controversies are slow to channel information to a wide public. When they succeed, the media tend to polarize the agents involved in arguments into two simple camps, the establishment versus the challengers. Professor Mazur reports that the greater the coverage of the technical controversy, the more negative public opinion becomes regarding the technology involved and the establishment position. This is illustrated in studies of nuclear power and fluoridation. One of the dilemmas of responsible reporting is that journalists lack the technical expertise to weigh conflicting scientific opinions and so are unable either to "referee" controversies or to sort out the factual from the policy issues. In this regard Professor Mazur argues in favor of the value of science courts.

In a spirited response, J. Richard Ponting raises several issues in the original paper, the most notable of which concerns the ability of a society, based on political action, to act rationally in the face of technological change. For Professor Ponting, rationality and politics do not make good bedfellows. He similarly speculates that little remedy will be found in science courts. These are sobering suggestions in view of our society's commitment both to democracy and to rational modes of action.

The second key paper deals not with the news, but with the entertainment industry. Dorothy Nelkin chronicles images of the atom in science fiction literature and movies from the turn of the century to the present day. Professor Nelkin suggests that this collective memory of atomic-driven fictional disasters may be instrumental in fueling the contemporary opposition to all-too-real nuclear weapons. In his counterpoint, Allan Olmsted explores a more mundane source of opposition—the dangers associated with ordinary nuclear technology, and the natural fears that develop from these dangers. The debate here is over whether what has been merely entertaining in the past will turn out to be relatively accurate *deja vu* for the future. The reader will be the judge as to who is ultimately correct in these discussions. Nonetheless, no one can dismiss the role of the media in amplifying our experience of technological dangers and our responses to them.

8

The Mass Media in Environmental Controversies

ALLAN MAZUR

The mass media have focused public attention on numerous environmental controversies in the United States and Canada, including those over nuclear power plants, acid rain, fluoridation, and various operations associated with the extraction and transportation of fuels. As diverse as these disputes are, they have shared features that make it convenient for the sociologist to lump them together. Typically a focal point of each controversy is a product or process of technology. Some of the principal participants in the dispute qualify as expert technologists or scientists. Almost invariably these experts appear on opposing sides of the controversy, disagreeing over relevant scientific arguments that are too complex for most laymen to follow. This is a perplexing feature of technological disputes, for while we are used to scientists contradicting one another on matters of policy, we usually expect them to agree about the facts of the physical world (unless one has made an error or lied), and when they do not, as happens so often, most of us are impotent to resolve the issue. This is a special dilemma for reporters in the news media, who rarely have technological training and yet must convey the substance of these arguments to a public that also lacks technical background.

In the United States, two distinct sides usually emerge in any environmental controversy that has escalated to the point of widespread public attention, and though one can often find more than two factions, these factions will be identifiable with one side or the other. It is usually convenient and accurate to label these the "establishment" position, and the "challenger" position which opposes the establishment. The establishment position is associated with some combination of government and corporate industry, while the challenger side is associated with voluntary organizations such as environmental and consumer groups, or *ad hoc* groups formed specifically to promote this protest or set of related protests. The challenger side may include some establishment figures and perhaps an occasional state or federal agency, but in these cases it is clear that they do not represent an establishment point of view. The situation is illustrated in Exhibit 1, which emphasizes the role of the mass media as an overseer of the controversy as well as a forum in which the debate is carried out before the public. The establishment-challenger model may or may not be appropriate for Canada, though it certainly applies in some cases (see Council on Environmental Quality, 1980). In any case, one clear difference between the American and Canadian situations is the strong American emphasis on litigation. For example, a regulatory agency may set a standard on, say, effluent of a carcinogen from a chemical plant. The agency is often sued, either by the plant, which may think the regulations too strict, or by environmentalists who consider them too lax. In either case, the suit goes to court where conflicting technical claims are heard about "safe" levels of population exposure to the carcinogen. Judges know little of technical matters and usually base their decisions on procedural matters, asking if the regulatory agency followed the proper form and acted within its mandate in setting the standard (Bazelon, 1979). Technical policy is thus set without much regard for the substantive issues in dispute. Environmental disputes seem to be "resolved" differently in Canada, either being left to the voluntary actions of developers, at the one extreme, or made the focus of full-blown public inquiries, at the other (Clark, 1981), though often on the middle ground there is informal negotiation between the regulatory agency and the offending industry (see Thompson, this volume).

THE MASS MEDIA

The mass media are the primary link between active participants in a controversy and the wider public. A few recognized spokespersons for the establishment and challenger sides supply most of the information reported by major national news outlets such as *The New York Times* and *The Washington Post, Time*, and *Newsweek*, the wire services,

Exhibit 1
Sterile Abstraction of Many Controversies

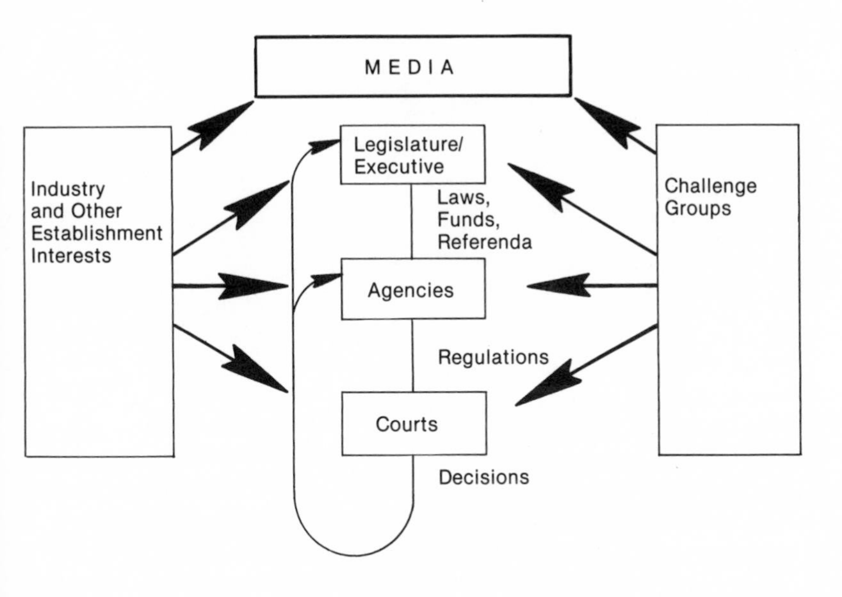

and television network news. Not many reporters specialize in environmental affairs; even these major national sources have only one or a few. The small number of major sources and the small number of major reporters come to know one another, sometimes on a friendly basis, so their personal relationships can affect the passage of information among them. The general public learns of the controversy, either directly from the national media or from local media which pick up and relay the national stories, sometimes supplemented by local sources (Exhibit 2). Thus, information flows to the public through a narrow channel that is regulated by a small number of activists and media people.

Since environmental reporters follow the technical literature for controversial items that may be newsworthy, we often see a sequence that goes like this: Item appearing in a technical publication is noticed by

Exhibit 2
Flow of Information from Activists to the Wider Public

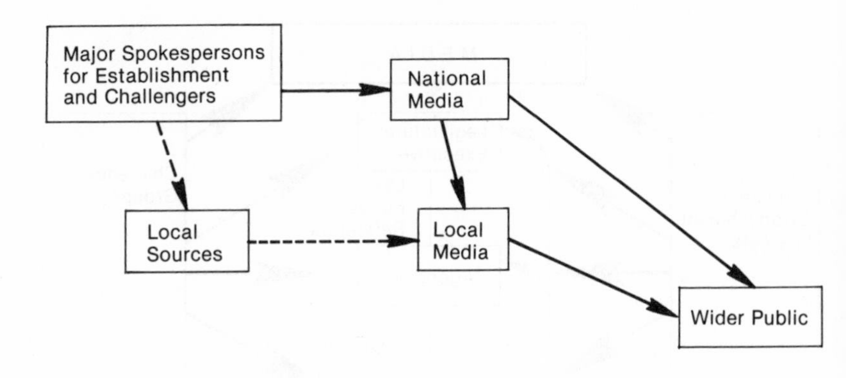

an environmental reporter, who then contacts the original author for more information. Sometimes the author initiates the contact, particularly if he or she knows the reporter. The reporter writes a story for the non-technical media, either a special interest (that is, environmental) publication or the national media. The local media pick up the item from the national media and relay it to their communities, perhaps supplementing it with local sources.

A good example is the acid rain issue whose movement can be followed through the magazines from citations in the *Readers' Guide to Periodical Literature*. Early technical reports on acid rain appeared during the early 1970s in the technical journals, *Chemistry* and *Science* (Exhibit 3), where they were picked up by the special interest magazine, *Environment*, and by the national news magazines, *Time* and *Newsweek*. Continued reports in the technical journals (and popular science periodicals) received, by about 1979, substantial coverage in the whole range of magazines catering to environmental and outdoors interests and in the general media as well, including *Macleans*, since the topic had a special relevance for Canada. Exhibit 4 shows the cumulative effect of these articles which, added to the newspaper and television coverage (not shown), brought acid rain to public attention as an environmental issue. Exhibit 4 shows a decrease in magazine coverage from 1980 to 1981, though there has been no alleviation of acid rain. Unfortunately the media are fickle in their choice of subjects, sometimes following fad cycles in allocating their limited space.

Air time for a half-hour television news program is preallocated to world, national, or local news, regular features, editorials, and advertisements. Newspapers and news magazines preallocate their page space in a similar way, with the responsibility for each section given to an editor and associated personnel. These are stable formats, matched to the production organization, and they are changed only in unusual circumstances. Thus, both print and electronic media have predetermined "news holes" that must be filled for each edition. The selection of those events of the day to be included or excluded, and the space and placement allocated to included items, reflects a reasonably objective appraisal by editors of which among these events are the most "newsworthy." Nonetheless, other factors obviously intervene, such as the "human interest" in a particular item. The preexisting working relationship between a news source and a reporter may also facilitate the placement of an item.

Three Mile Island was obviously a newsworthy event, yet the quantity of coverage given to it far surpassed that given two months later to the worst American airline accident in history, the crash of a DC-10 killing 274 people. Among the factors contributing to the inordinate quantity of coverage were these: the power plant accident occurred at a time of public concern with nuclear power, so media people were sensitized to the issue and were easily attracted to the event. By a quirk of timing, the anti-nuclear movie *The China Syndrome* was released to theaters across the nation just days before the accident. The event itself had great human interest—the week-long struggle with the bubble, the heroics of the nuclear engineers in averting disaster, the entry of the President, the exit of the frightened populace: all this had the drama of a soap opera, to be followed day after day. In addition, the plant site in Pennyslvania was easily accessible to reporters in nearby New York and Washington (Sandman and Paden, 1979). Each of these factors encouraged high coverage.

MEDIA COVERAGE AFFECTS THE WIDER PUBLIC

The nation becomes a spectator to an environmental controversy through the press and television, particularly during those periods when

Exhibit 3
"Acid Rain" Magazine Articles Indexed in *Readers' Guide*, 1972-81

1972	1973	1974	1975	1976	1977	1978	1979	1980	1981
Science									
Chemistry		*Chemistry*			*Chemistry*				
		Science	*Science/2*	*Science*			*Science/3*	*Science/4*	*Science/5*
		Sci. News				*Sci. News*	*Sci. News*	*Sci. News/4*	
		Sci. Dig.					*Sci. Dig.*		
						BioSci.		*BioSci./2*	*BioSci.*
							Sci. Amer.	*Sci. Amer.*	
									Sci. Quest
Environment, Outdoors									
Envir.						*Envir.*	*Envir.*	*Envir./2*	
				Horticult.					
						Nat. Pk.	*Nat. Pk.*	*Nat. Pk./2*	
							Conservat.		*Conservat.*
							Field & Stre.		
							OutdoorLfe		
							Blair&Ket.		
							Org.Gard.		
								Sierra	*Sierra*
								Int.Wldlfe	
								Audubon	
									Nat. Hist.
									Nat. Geog.
General									
Newsweek							*Newsweek*		
		Time						*Time*	
						Sat. Rev.			
							US News&WR		
								Macleans	*Macleans/4*
								Read.Dig.	
								Forbes	
									Sports Illus.
Specialized									
							Smithsonian		
							Futurist		
							Change		

Exhibit 4
Number of "Acid Rain" Articles in *Readers' Guide*, by Year

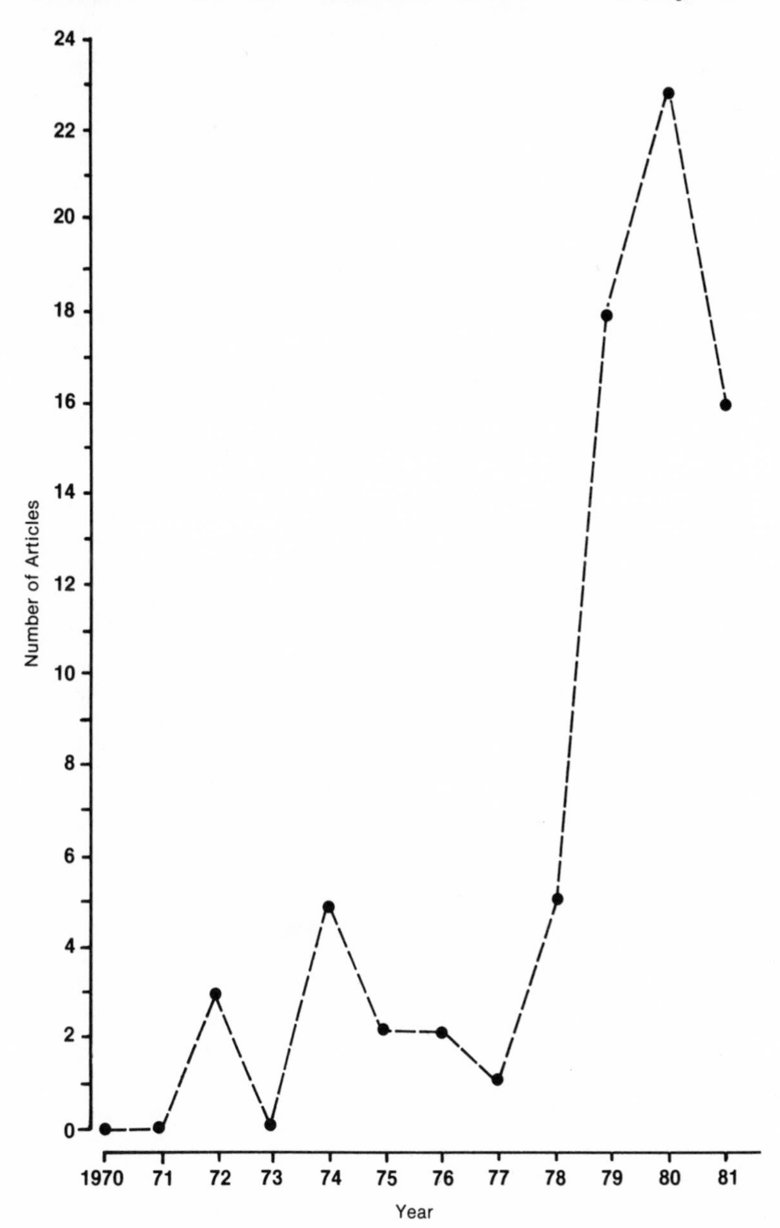

protest activity and media coverage are intense. I have studied in detail the controversies over nuclear power and fluoridation, in both cases the protests opposing implementation of these technologies. Judging from these cases, as the quantity of media coverage increases, public attitudes toward the technology (as measured in opinion polls) become increasingly negative. This is illustrated in Exhibit 5, which compares yearly fluctuations in magazine coverage of nuclear power with public opposition (Canadian and American) to nuclear power plants. The trends show corresponding peaks and valleys. A similar pattern for American opinion appears in the fluoridation controversy (Exhibit 6).

These trend data are crude. The accident at Three Mile Island allows a finer view of the link between media coverage and public opinion because, in the year following the accident, the Harris Poll took numerous opinion surveys in the United States closely spaced in time (Mitchell, 1979). This fine-grained opinion trend can be compared to (smoothed) weekly fluctuations in coverage of the accident on television network news, in *The New York Times*, and in the major news magazines as shown in Exhibit 7.

The accident began on March 28, 1979, and in the week following, almost 40 percent of television-network evening news was devoted to it. Coverage in the news magazines was necessarily delayed by one week, but both *Time* and *Newsweek* ran cover stories in April. By June the story had disappeared from the news magazines and appeared in only occasional short pieces on television until about October, when there was a second, much smaller, rise in coverage to report the final work of the Kemeny Commission which had been appointed by President Jimmy Carter to investigate the accident. The Commission's report was released at the end of October, but the media had been anticipating it, reporting related events throughout October.

The proportion of the public opposing the building of more nuclear power plants rose sharply after the first burst of coverage. This is hardly surprising since the spectre of the accident would be expected to increase opposition regardless of any independent effect of the quantity of coverage. However, one would not expect, on this basis alone, that support for nuclear power would rebound within two months, as soon as the media coverage had decreased; yet that is what happened in both Canada and the United States (see Exhibit 5). Furthermore, a clear, short-term increase in American public opposition appeared again during October and November, with rebound by December, coinciding

Exhibit 5
The Nuclear Power Plant Controversy

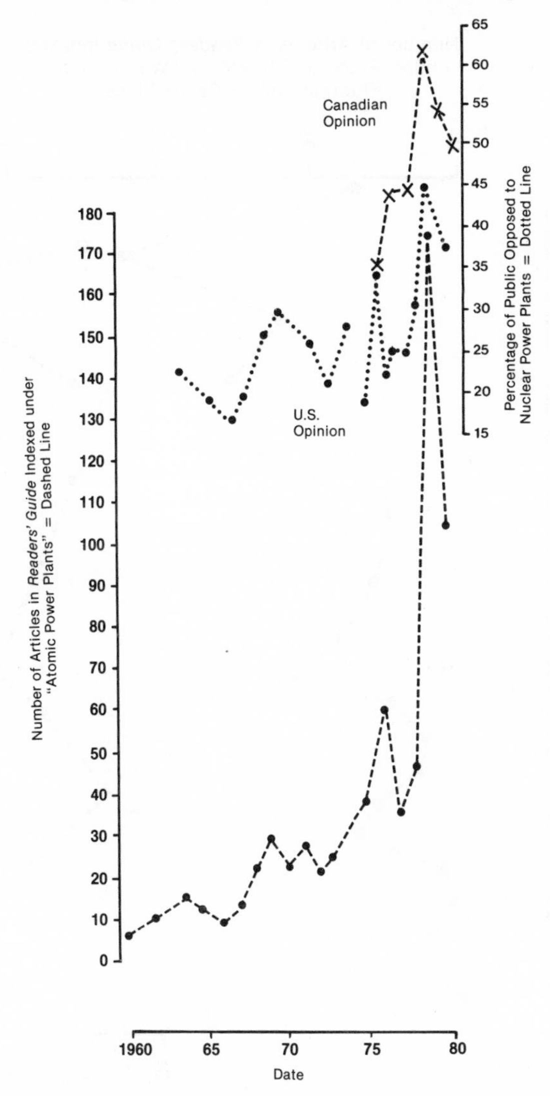

Note: The American opinion data are described in Mazur (1981a: 117). The trend line is broken at 1975 because the question before that date differs from later wording. The Canadian question, taken by Gallup, asks, "At present, very little of the total electricity used in Canada comes from nuclear power generation. What do you think should happen?" The responses, "They should not develop any more than at present," and "They should stop generation of nuclear power," were combined to represent opposition.

Exhibit 6
The Fluoridation Controversy

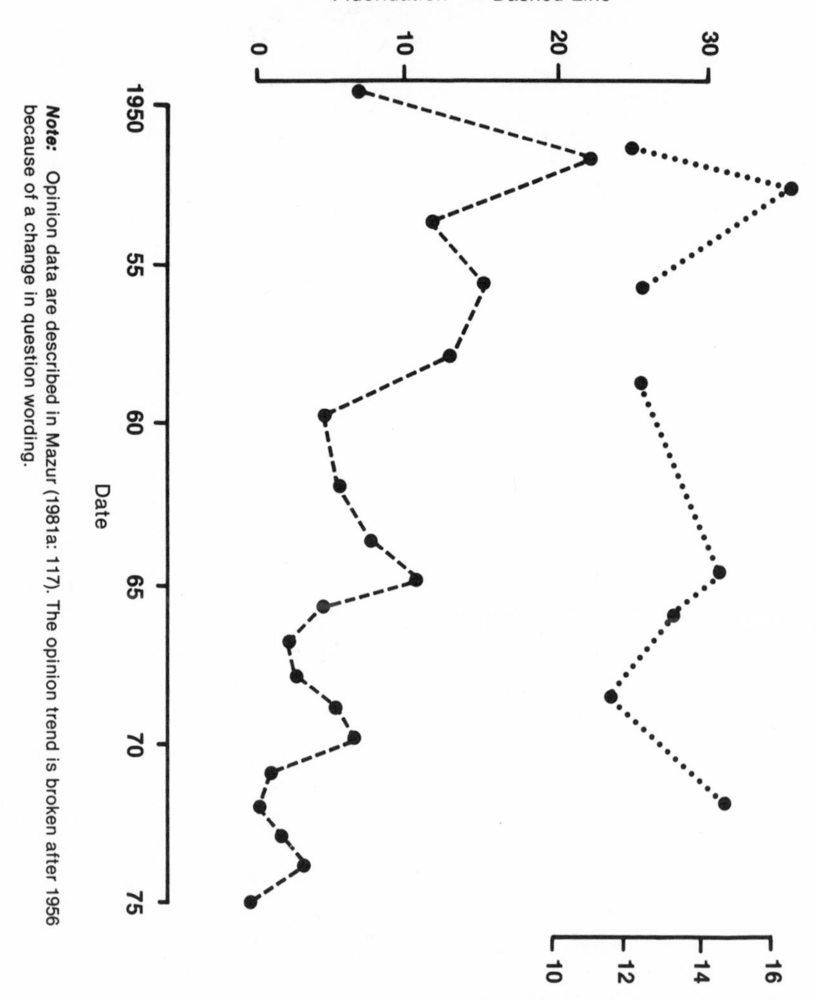

Note: Opinion data are described in Mazur (1981a: 117). The opinion trend is broken after 1956 because of a change in question wording.

Exhibit 7
Public Opinion Toward Nuclear Power Plants and Media Coverage of the Three Mile Island Accident

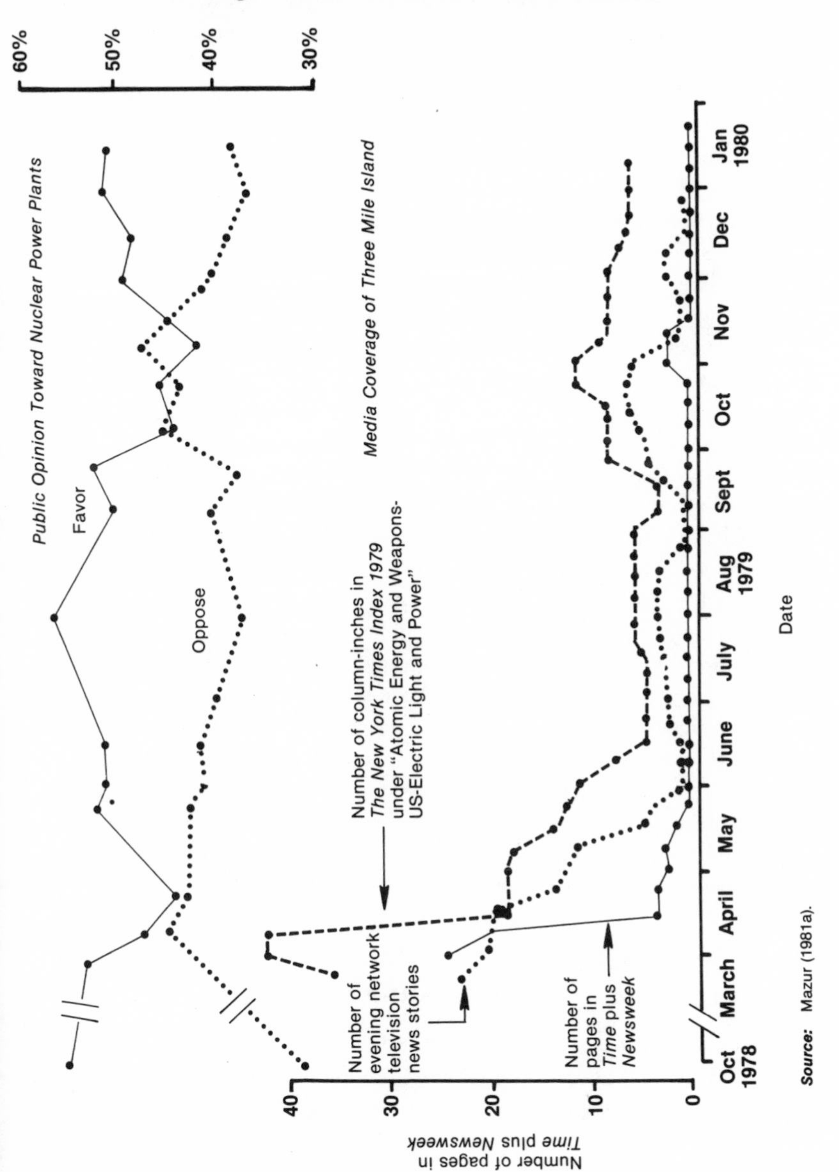

Source: Mazur (1981a).

perfectly with the secondary peak of media coverage at the time of the Kemeny Commission's final work. The Canadian opinion polls are too spaced out in time to test for this effect. The parallelism between these trends supports my contention that public opposition is a function of the quantity of media coverage.

Why does public opposition increase as media coverage increases, even in cases such as fluoridation where the media do not carry an obvious bias against the technology? Perhaps the prominence given to disputes between technical experts over the risks of the technology makes it appear dangerous to the public. Persons experimentally exposed to both positive and negative arguments about fluoridation were more likely to oppose it than persons who had not seen any of the arguments at all (Mueller, 1968). Fluoridation is more likely to be defeated in a referendum when there has been heated debate than when the campaign has been relatively quiet (Crain et al., 1969). During local controversy over a proposed nuclear waste storage facility, residents who had heard or read about the controversy were more likely to oppose the facility, and to consider it unsafe, than residents who had not heard about the controversy (Mazur and Conant, 1978). If doubt is raised about the safety of the technology, many in the public prefer to err on the side of safety, as if saying, "When in doubt, reject the technology—better safe than sorry" (Sapolsky, 1968).

My simple characterization of public opinion ignores the varying responses of different population groups to the media. The attentive public responds differently from those who are less interested in the media. People who already have strong opinions regarding an environmental issue are less swayed by the media, in part because they selectively expose themselves to concordant opinions, and they are more likely to believe coverage that supports their prior views. Of course, the influence of the media interacts importantly with the reactions of one's friends and associates to the items being reported.

THE DILEMMA OF RESPONSIBLE COVERAGE

In simpler times the news media could be viewed as a passive reporter of daily events, as long as journalists' opinions were restricted to editorials. Increasingly we have come to see the media as an active shaper of the events that are reported. The press and television have been full-fledged participants in the nuclear power controversy. Could they be anything else?

If public attitudes in an environmental controversy are influenced by the simple quantity of media coverage, how does one report the controversy without at the same time affecting it? Certainly there are instances in which one could argue whether or not particular journalistic practices were improper (Rubin et al., 1979), but the dilemma is deeper than that. My own recent study of newspaper reporting in the fluoridation controversy shows that even an ostensibly objective, even-handed story about the pros and cons of the technology tends to shift readers toward opposition (Mazur, 1981b).

Environmental controversies are chaotic. Each side is ultra motivated to make the strongest case it can while at the same time finding flaws in the other side's position. Unfortunately, in most controversies the adversaries never confront one another. Instead they address different audiences or the same audience at different times or in formats that allow them to speak past one another. It falls to the media reporter to sort the arguments, juxtaposing one against the other, often in apparent contradiction. We become confused when one expert seems to disagree with another about basic facts, and we cannot locate the reason for their disagreement. The reporter rarely has the skill or credentials to pass credible judgment over the conflicting claims, so he or she usually presents them as equivalent, even when one claim is held by only a single dissenting expert. It is not surprising, in these confusing circumstances, that laymen adopt a defensive stance toward the controversy, preferring to reject it rather than confront the swirl of contradictory claims. If we could bring some order to the chaos, we might improve the factual basis for public decision-making, and we might then improve the decision-making itself.

Some observers see benefit in separating disagreements over fact from disagreements over value (or policy) questions. Factual disputes may be resolved by the analysis of data and so can be treated differently from value disputes that are impervious to the methods of science. For example, in the debate over nuclear power plants, scientific data could tell us whether or not X cases of cancer are likely to occur in a population exposed to Y amount of radiation, but they cannot tell us if X cases of cancer are an acceptable price to pay for the electricity that would be generated by such a power plant.

This separation corresponds nicely with the division of labor in our society between scientists and policy-makers. The scientists have the skills to resolve factual disagreements, or at least to understand the basis for disagreement, but they have neither special skills nor the public mandate to make policy decisions. The policy-makers, either the full electorate or their delegated authorities, have the mandate to make policy but usually lack the technical skills to deal with ambiguous scientific questions. If somehow the scientists could resolve factual disagreements, then this resolution could be reported in the media as a consistent and credible statement of fact, which could then serve as a firm basis for debate over desired policy.

THE SCIENCE COURT

The "scientific court" is one approach to the resolution of factual disputes after they have been separated from the value differences that are usually meshed with them (Task Force, 1976; Mazur, 1977). Controversial issues would be referred to a science court by a means not yet specified, but perhaps through a request from a legislature or a public inquiry board. The procedure would begin with the selection of adversaries to represent both sides. The adversaries would be asked to state the scientific facts they considered most important to their respective cases, and to supply documentation to support their assertions. They would then exchange their lists and documentation, examining each other's claims and specifying those assertions with which they agreed and those with which they did not. A referee who was acceptable to both adversaries would attempt to arbitrate differences between them, perhaps by alterations of wording or by removal of ambiguities in factual statements. In the event both adversaries agreed on all listed statements of fact, the procedure would end, and these facts would constitute the science court's report.

If one or both sides challenged factual statements by the other, these challenges would be the subjects of a hearing, open to the public and governed by a disinterested referee, in which the adversaries argued their opposing scientific positions before a panel of scientist-judges. The referee would restrict the debate to questions of fact, excluding issues of moral judgment and policy. The judges themselves would be established experts in fields relevant to the dispute, but would not include anyone with a personal bias or organizational affiliation that would predispose toward one side or the other. The judges would be subject to challenges by the adversaries on the grounds of prejudice, and the court would only proceed once a panel of judges acceptable to both sides had been selected.

After the scientific evidence had been presented, questioned, and defended, the judges would prepare a report on the dispute, noting points on which the adversaries agreed and reaching judgments on factual issues still in dispute. The judges might be able to decide if either or both of the adversaries were wrong, or if the differences between them were legitimate, resting on points of irreducible ambiguity or insufficient data. They might suggest specific new research to clarify points that remained unsettled.

It is important to emphasize that the judges would not make a recommendation on policy or a judgment of moral right or wrong. Scientific judges are assumed to have no special wisdom that permits them to decide moral issues in their society. They should, however, be particularly qualified to arrive at purely scientific conclusions, and this would be the limit of their mandate. The court's report could then serve as the factual basis for value judgments by other bodies more properly charged with policy-making functions, such as the legislature or a public inquiry board or the wider public in the case of a referendum.

9

Counterpoint: Scientific Debate and Politics in the Science Court

J. RICHARD PONTING

Allan Mazur's analysis of the role of the mass media in environmental controversies, while interesting, can be improved upon. Mazur appears to vacillate over whether the "environmental controversies" of which he writes are a distinct phenomenon in and of themselves, or whether they are merely a special case of a broader category of "technological disputes," as the initial version of the paper labeled them. The lead paragraph suggests an emphasis on the environment, but the concluding section on "the science court" suggests an emphasis on technology. If the latter is to be our guide, we are left in need of a more theoretically sound definition of what constitutes a "technological dispute." Does the highly contentious issue of the accuracy of public opinion polls which measured interest in separatism in Quebec prior to the 1980 referendum constitute a "technological dispute" to be referred to the "science court"? As it turned out, the conflicting results of the various pollsters were referred to a panel of social science methodologists for resolution. We need a better grasp of the scope of technological controversies.

In responding to my initial criticism of his dichotomy between "establishment" and "challenger" positions on an issue, Professor Mazur suggests that "the challenge side may include some establishment figures and perhaps an occasional state or federal agency, but in these cases it is clear that they do not represent an establishment point of view." Notwithstanding Mazur's hedging on whether the "establishment-challenger" dichotomy has general applicability to Canada, it is important to stress here the notion of countervailing powers. The American political system itself has institutionalized countervalence through its elaborate system of checks and balances which often result in the challenge position issuing forth from what is merely another branch of "the establishment" (for example, a challenge to the White House from the Congress). In Canada the philosophy of countervalence was dear to the heart of former Prime Minister Pierre Trudeau. Lacking constitutionally entrenched checks and balances, he often deliberately created them within the federal bureaucracy itself or directed state funds to external organizations to serve the same function. When these countervailing sources of power are located in the state bureaucracy itself, it may be naive to overlook the establishment origins of the challenger position.

The establishment-challenger dichotomy is questionable in other instances as when, for example, state regulatory agencies such as Alberta's Public Utilities Board and others require the proponents of a project to fund interventions in regulatory hearings by those who are challenging the project. The simple establishment-challenger dichotomy also seems to overlook the kind of situation which Richard Schultz documents in his book *Federalism, Bureaucracy, and Public Policy* (1980), where governments actively lobby the lobby groups! These points require a more complex analysis of the establishment versus challenger situation.

A more serious question concerns the role of the media. Mazur points out that the media cannot "report the controversy without at the same time affecting it." Yet he overlooks a large body of literature in the sociology of the mass media (for example, Molotch and Lester, 1974; Sigelman, 1973; Spector and Kitsuse, 1977; Tuchman, 1978) and the sociology of knowledge which points to the *social construction of reality*. This would seem most relevant to his concerns. Of course, the media are going to affect issues; indeed, they are going to affect the very definition of what the issue is. Issues do not exist on their own, floating around in the air just waiting for us to recognize them. Issues, including environmental "technological" issues, are human creations manufactured by the social actors (for example, the moral entrepreneurs, the career-enhancing journalists, the self-interested government bureaucrats) depicted so well by Spector and Kitsuse in their model of social problems as claims-making and claims-responding activities.

In response to an earlier suggestion, Mazur claims that different segments of "the public" will respond differently to mass media coverage of an issue. For instance, the response of the "attentive public" will be different from that of the "mass public." What needs to be explicitly recognized here is that publics expand and contract and that people may be pulled into lower (more central) "orbits" of a public in times of controversy and may drift away to become peripheral members of that issue's public at other times. Because negative media coverage of nuclear power is likely to cluster in periods of accident and controversy, and because positive coverage is most likely to be found during other periods, we may be dealing here with different publics—namely, one that attends to both positive and negative coverage, and one that mainly attends to the less favorable crisis coverage. This helps to explain the apparent volatility of public opinion on nuclear power which Mazur plots in Exhibit 5.

Furthermore, it is an open question whether the sheer volume of media coverage that appears to have a reactive effect on public opinion polls actually affects the public policy decision-makers. In the first place, the opinions of the mass public can at most set very broad parameters within which policy decision-making will occur; public opinion can never be informed enough in today's complex world to influence decision-making on the myriad of "technical" sub-issues encompassed within a larger issue. Policy-makers will be particularly skeptical of data that show only the direction of public opinion on the issue at various points in time, and not such other fundamental characteristics as the *intensity* with which the opinions are held, the extent of polarization which the opinions represent, and the depth of knowledge on which the opinions are based.

Another question to be explored is the extent to which public opinion data can be placed in a real-world context. Public opinion gathered in a vacuum, which excludes from respondents' consideration the trade-offs that have to be made with various technological choices, must be of limited utility. As our colleague, Roger Gibbins, explains it: respondents who are asked about energy issues in Alberta need to be asked *how* they want their power—with strip mines in the southern Alberta coal region, with leaks and meltdowns of nuclear reactors in central Alberta, with river pollution in the tar sands region, with offshore oil spills in the Beaufort Sea, or with acid rain in Saskatchewan from gas plant sulphur dioxide emissions in central Alberta. The same dilemma applies elsewhere.

Perhaps the most serious question to be raised in Mazur's approach is the implicit adoption of the controversial rational planning or rational-scientific model of public policy-making and his optimism regarding the neutrality of the science court. (See Scott and Shore, 1979, on why sociology does not apply to policy.) The rational planning model of public policy assumes that policy-making proceeds through a scenario whereby policy-makers "gather all the facts" surrounding a problem, identify alternative policies for responding to that problem, identify the criteria for selecting among competing policy options, and then make their choice by assessing the options in light of the criteria. The model optimistically assumes not only that decision-makers have the time at their disposal to go through this process, but also that this is the ideal way of making public policy in that it allows no role for crass political interference or emotionalism.

The major shortcoming overlooked by the rational planning model is that the formulation of policy problems, whether for a bureaucrat in a nuclear regulatory agency or a panel of scientific experts in Mazur's "truth tribunal," is inherently a political phenomenon. The scientific method does not lend itself well to choosing between two competing formulations of a problem, because problem formulation is itself a value-laden activity. It is part of the essence of politics. Thus, to presume that what will be referred to a science court will be pristine "scientific" problems, stripped of value baggage, would not be realistic. This observation is well established in view of current developments in the philosophy of science (see Kuhn, 1970; Feyerabend, 1975).

Even in the unlikely event that problems referred to the science court could be formulated in value neutral terms, serious problems would arise in the selection of adversaries to represent the respective sides. Who would do the selecting of the adversaries? The experience of editors of academic journals in selecting referees for articles might be instructive here. As journal editors will attest, the best persons to approach to referee any given article are often not at all self-evident, and editors seldom reach a consensus on whom to approach. The same problem arises with the selection of the scientific judges. Furthermore, both the judges and those who select the judges and the advocates will likely approach the task wedded to a certain scientific paradigm. In short, contrary to what Professor Mazur seems to envisage, they are unlikely to be a "blank slate" free of their own biases. Indeed, it would be erroneous to view the present courts, such as the Supreme Court in Canada or the United States, as neutral in policy-making.

In real life the model of a prestigious panel of neutral scientific judges would likely be short-lived, if it ever got off the ground at all. The proceedings themselves would likely be demeaned as academics (in whatever role) came to use the court as a forum for prestige-seeking, disciplinary, argumentation, and "grandstanding" or "showmanship." Similarly unless protected by contempt of court provisions in law, the rulings of the scientific judges would likely be discredited on ideological grounds. This has happened frequently in jurisprudence where important legal decisions have been viewed as ideologically motivated. This has occurred, for example, in Canada, with respect to Mr. Justice Emmett Hall's pronouncements on the danger which physicians' extra-billing poses to universal medicare in Canada, and with Mr. Justice Thomas Berger's decision on the dangers of a Mackenzie Valley pipeline in the far North. In the scientific court proposed by Mazur, the scientific professions will have their clay feet exposed, and those professions, in the public spotlight like never before, might quickly lose their mystique and with it the legitimacy which is the cornerstone of Mazur's scientific court.

Mazur recognizes that factual issues might remain in dispute at the end of the court hearing and that judgments will have to be rendered on these. Indeed, one might suggest that the facts will almost never be incontrovertible, for otherwise the issue would be unlikely to be referred to the scientific court at all. Where there remains irreducible ambiguity or irreconcilable interpretations of "fact," the paradigmatic and other value biases of the panel will inevitably enter. Alternatively, the scientific court can render an inconclusive verdict or, as Mazur notes, suggest specific new research to clarify the unsettled points. This will reinforce stereotypes of the fence-sitting academic and will challenge the legitimacy of the scientific court in the eyes of both the public and the politicians. The politicians, who have the power and authority to create and dissolve the scientific court as an institution, will be particularly dissatisfied, for they must be *seen* by their constituents to be doing something about societal problems, rather than waiting patiently while academics and other scientists dither or dally and engage in debate which to the lay person may be unintelligible sophistry.

Consequently, in the opinion of this observer, the proper place for such scentific debate is in the meeting halls of the learned societies and in the pages of the learned journals, not in the chambers of a new scientific bureaucracy which the science court would quickly become.

10

Popular Images of the Atom and Mobilization of the Anti-Nuclear Movement

DOROTHY NELKIN

The anti-nuclear movement has persisted in the face of frustration, repudiation, and cooptation. It has also succeeded in attracting a very broad constituency, gaining momentum over more than a decade as a significant political force. How does one understand the persistence and broad appeal of this social movement? Recent perspectives on social movements focus on the importance of their organization, and the process by which they develop their social base (McCarthy and Zald, 1977). Much of this literature emphasizes the incentives for the individuals who participate (Fireman and Gamson, 1977). Rather less has been written on the means by which shared consciousness actually develops over a particular issue or goal (Feree and Miller, 1980).

This paper uses the nuclear debate as an example of how images can serve as a mobilizing force. Political discourse is intended to persuade and does not necessarily seek analytic clarity. The use of metaphor in political discourse serves as a strategic tool, a means to construct a common scheme of reference, a way, in other words, through which activists recruit and mobilize followers, establishing a vision of reality that appeals to diverse groups.

The argument of this paper is that the growth of the anti-nuclear movement and its broad appeal can be attributed to the symbolic char-

This paper is part of a project supported by the National Science Foundation's EVIST Program. The research assistance of Lauren Stefanelli, and the criticism of Spencer Weart, John Woodcock, and Rose Goldsen are greatly appreciated.

acter of nuclear power, its apocalyptic image, and its association with war. This image has survived the linguistic tactics to dissociate the civilian from military uses of atomic energy. In the 1950s, for example, "atomic" power became "nuclear" power, nuclear industrial areas became "parks," and accidents became "incidents." Labels failed, however, to erase deeply rooted images. Similarly, documentation of the safety record of the nuclear industry failed to dispel public fears. While chemical plants, liquid natural gas storage facilities, and toxic waste disposal sites arouse local concerns, no other technology has evoked such devastating visions, such apocalyptic metaphors, as nuclear power.

This remarkable aspect of the anti-nuclear discourse has contributed to the persistence and depth of public concern about this technology. Activists use powerful and macabre metaphors to describe the technology. Buttons, bumperstickers, books, and balloons convey the mood of the anti-nuclear movement. They portray nuclear power as "a silent bomb," "a berserk technology," "poisoned power," "nuclear madness," "a game of genetic roulette." Slogans play on anxiety and fear. "The question is one of survival," "nuclear war by 1999," "there's no future in nuclear power," "ain't nowhere we can run." A gallows humor prevails. "Mutations are America's hope for the future," "Population control by atomic pollution," "Better active today than radioactive tomorrow." Proposals to recycle nuclear waste include coffee heating crystals which will perk up temperatures to a "hearty 1000 degrees." A company called Last Blast Endeavours puts out a Christmas gift, a radioactive waste container fully equipped with a melt top plug, a disaster crate, and simulated radioactive waste that glows in the dark. An operator's handbook notes that with care it can last half a lifetime. Nuclear games include "Plutonium Tag" in which the person tagged is contaminated and must indicate it by sizzling and glowing in the dark.

Anti-nuclear graphics portray similar themes. Matched buttons show a cow saying "no nukes" and a nuclear plant saying "no cows." Posters show skeletons and mutants. Mushroom clouds come from cooling towers. A worker hammers nails into an endless row of coffins on a poster saying "nuclear power creates jobs." An official stands on a pile of burnt out rubble that was once a utility: "I repeat there is no cause for alarm." A reactor shaped like an egg begins to crack, giving birth to a bomb.

Although nuclear weapons are seldom an explicit target for the anti-nuclear movement, the images of war, death, and extinction dominate its discourse. The power of this imagery has been a source of unity for the heterogeneous constituency of the movement, and activists effectively use macabre metaphors to broaden their public support. Indeed, despite continued efforts to distinguish the images, most people continue to associate nuclear power with war. A survey by the Council for Environmental Quality asked, "Is it possible for a nuclear power plant to explode and cause a mushroom shaped cloud like the one at Hiroshima?" Fifty-two percent of the respondents said yes or maybe, and 16 percent did not know. Only 28 percent would accept the siting of a nuclear power plant within 14 miles of their home, while 63 percent would accept a coal-fired plant within the same distance (Council on Environmental Quality, 1980).

Why is this apocalyptic image so pervasive? What is it about nuclear power as compared to other potentially devastating technologies that evokes such a visceral response? Science fiction and films provide some helpful clues to the contours of public opinion. Such works of the imagination are shaped by the attitudes and concerns, and even the distortions and misconceptions, of both their producers and those who sustain them through box office returns. But the metaphors in popular films and fiction also help to create the symbolic environment that shapes our values and our beliefs. The atom has been an especially potent metaphor, but more than a metaphor the image of the atom is packed with messages about the human control of technology, about the social and political implications of a "nuclear society."

This paper describes the spectrum of meanings that have developed around the concept of the atom as expressed in popular culture. Three themes recur in many forms. First, the atom in fiction portrays the devastating destruction of humankind through our own technology; annihilation of the species is a persistent theme. Second, the atom in fiction carries the image of mutation, or transmutation with all its mystical overtones. Third, fiction and films have used the atom to represent the irreversibility of technology amenable to control through neither political negotiation nor human effort. The image of the atom is one of technology-out-of-control.

THE ATOM IN POPULAR CULTURE

The atom has exercised the imagination since the days of the early discovery of radioactivity. Historian Spencer Weart described the emotion and anxiety surrounding Ernest Rutherford's and Frederick Soddy's revolutionary discovery in 1901 that radioactivity indicates atoms changing from one element to another (Weart, 1980). From that time on the atom acquired an ominous fascination in popular culture, providing both a new form of fantasy and a fantastic form of news. Science writers and Hollywood producers exploited the atom as a source of entertainment well before its performance in Japan brought its dramatic potential to the attention of newspaper readers. In a serial published in 1913–14, *The World Set Free*, H. G. Wells wrote about a devastating war in the mid-twentieth century marked by explosions caused by small atomic bombs that set off a "furious radiation of energy, and nothing could arrest it. . . . The atomic bomb had dwarfed the international issues to complete insignificance." In 1918, another science fiction writer, Doc Smith, cited uranium as a rocket fuel in his Skylark series. An early Heinlein story in 1940 focuses on the quality of the fission process that makes it difficult to use as compared to conventional operating technology and in particular suggests the psychological pressures faced by workers who fear catastrophic accidents as nuclear energy replaces oil.

During the 1930s, monster and mad scientist films began to exploit the atom. In *The Invisible Ray* (1936) a mad scientist, Boris Karloff, discovered radium X and was contaminated with radioactivity so that anything he touched immediately died. After a murderous rampage his body degenerated into a blazing inferno. Several years later in *Dr. Cyclops* (1940) a mad doctor invented a radium device that shrunk its victims to thirteen inches. Radium-based technology provided a new gimmick for classic adventure films; *The Flash Gordon* serial (1936) included the tortures of a "horrible atomic furnace." In a western serial *The Phantom Empire* (1935) the cowboy hero, Gene Autry, lived near a mysterious underground kingdom with such amenities as radium generators and a radium revival room to bring the dead back to life.

The "atomic machines" imagined by science fiction writers during this prewar period were not very different from the bolt of electricity that brought the Frankenstein monster to life—pure fantasy, safely distant from the real world. But they tried to inject a note of urgency in their work. In his foreword to *The Invisible Ray* (1936), producer Carl Laemmle wrote, "That which you are now to see is whispered in the cloisters of science. Tomorrow these theories may startle the world as fact" (Menville, 1975: 53).

After Hiroshima several studios tried to produce documentary-style films that were intended to inform citizens about the consequences of the development of the atom (Shaheen, 1978). But the first effort, MGM's *The Beginning of the End* (1946) documenting the development of the bomb, was reviewed as grossly inaccurate, overly romantic, and excessively sentimental. A second effort to tell the story in a pseudo-documentary style, *Above and Beyond* (1953), showed the psychological struggles of the pilots of the "Enola Gay," the plane that delivered the atomic bomb to Hiroshima. This, too, was panned as a "conglomeration of romantics and melodrama" (Shaheen, 1978: 9).

Indeed, in the 1950s the atom inspired what film critic Douglas Menville called a "renaissance" in the science fiction film (Menville, 1975: 74). Producers created a new genre of film focusing mainly on transmutation. In *Bride of the Monster* (1956) Bella Lugosi uses atomic power to transform human beings into a super race but instead creates a monster. The scientist in an appetizing film called *The Fly* (1958) makes an atomic machine to transform matter and used himself as an experimental subject. Unnoticed, a fly gets into the apparatus and the scientist ends up with a fly's head while the fly gets a human head. An army officer in *The Amazing Colossal Man* (1957) grows into a deranged giant after exposure during a plutonium blast. Lon Chaney, Jr., is mutated in *The Atomic Monster* (1953) and in *The Indestructible Man* (1956). A man turns into a monstrous lizard in *The Hideous Sun Demon* (1956). And, in a remarkable series of atomic creature features in the 1950s, ants (*Them*), wasps (*Monster from Green Hell*), moths (*Moths*), spiders (*Tarantula*), grasshoppers (*The Beginning of the End*), scorpions (*The Black Scorpion*), and an octopus (*It Came From Beneath the Sea*), all grow into the size of 747s after straying into the path of atomic tests. Soon, a totally different type of radioactively induced monster—an amorphous blob of radioactive matter on a rampage—appeared in *The Blob* and in *X, The Unknown*. Parenthetically, it is of some interest that when *Them* was issued, Twentieth Century magazine denounced it as a vicious allegory calling for the extermination, not of giant ants, but of Communists (Parish and Pitts, 1977: 318). This reaction does not seem to have arisen with the other films.

Classic science fiction themes of outer space, alien invasions, and future worlds also adopted the ominous imagery of the atom. In some cases the atom appeared as a useful weapon in the hands of heroes. Strontium 90 killed the monster in the *Island of Terror* (1956), and aerial dusting with atomic particles did in the enemy alien in *Kronos* (1957). Walt Disney's 1954 version of Jules Verne's *20,000 Leagues Beneath the Sea* turned Captain Nemo's submarine into an atomic-powered vessel, and *The Fantastic Voyage* (1966) created a miniature atomic submarine and crew that was able to travel through the human body. But most science fiction films still villified the atom as a sensational and uncontrollable means of destruction.

The science fiction films of the 1950s and 1960s were popular and probably cathartic, for their extraordinary level of fantasy allowed the observers to dissociate themselves from the events. But many of them contained explicit messages as well as metaphors. In the early days of the nuclear arms race—an era of backyard bombshelters, civil defense handbooks, and an arms race still new enough to be news—films began to focus on the danger of nuclear war. They conveyed explicit warnings: in *The Day the Earth Stood Still* (1951) Michael Rennie, as "an allegory of Christ," appeared in Washington, D.C., to warn the earthlings that atomic weapons testing must cease. Warning messages came from aliens from outer space who tried to save the earth (*The Space Children*, 1958), or from space travelers who found planets devastated by nuclear war (*Rocket Ship XM*, 1950) or even from "heavenly judges" (*The Flight That Disappeared*, 1951). In *Seven Days Till Noon* (1950) a scientist steals a nuclear bomb and holds a city for ransom in the vain hope of banning the bomb.

Some films explicitly pictured in devastating detail the consequences of nuclear war. Horribly deformed mutants roam in a "forbidden zone" while survivors live in a subterranean, *1984*–like society. This formula is repeated again and again in films such as *World Without End* (1956), *The Time Machine* (1960), and *Terror from the Year 5,000* (1958). Melodramatic and pessimistic to an extreme, most films of this type created vivid images but avoided serious treatment of the political issues leading to nuclear war; while labeling "heroes" and "villains," they avoided placing any blame. As one reviewer wrote of *Five* (1951), a futuristic vision of the world after the bomb, "the only function of (the) nuclear catastrophe appears to be an arrangement of the hero to get the heroine to fall into his arms" (Ernest Martin in Shaheen, 1978: 10).

A very popular film, *On the Beach*, was released in 1959 during the peak of the Cold War when nuclear war seemed a real possibility. It depicted the Australians' tortured wait for death as the radioactivity that had consumed the Northern Hemisphere descended inexorably toward them. It conveyed a powerful theme—the helplessness of humankind as nuclear war wipes out civilization. However, the producers avoided scenes that would outrage the audience so that a critic observed "we do not come to feel that we are in the hands of someone who cares deeply for our civilization" (Young, 1962: 95). Significantly *On the Beach* suggested that war can be attributed directly to the technology: as one protagonist observes, "the world was probably destroyed by a handful of vacuum tubes and transistors—but not by any policy or country or person." Tempered neither by political analysis nor by critical judgment, it conveys the image of technology beyond all human control.

In the less tense Cold War atmosphere of the mid-1960s, several films began to speculate on the cause of nuclear war. *Dr. Strangelove* (1964), *Fail Safe*, (1964), and *The Bedford Incident* (1965), all suggested that humanity is caught in its own deadly technology, that our culture is fundamentally marked by metaphor, and that simple human error or mechanical breakdown could irretrievably set off the machinery to end the world. Nuclear technology became a vehicle to express such themes. Again the dominant image is technological inevitability. No one is to blame; it is the technology itself, more than the decisions about it, that is responsible for devastating events. The three films all share this technological determinism, but it is most explicitly expressed in *Fail Safe* when the President of the United States commiserates with a Russian Premier about a weapon that neither can control:

> "This crisis of ours—this accident as you say. . . . In one way it's no man's fault. No human being made any mistake and there's no point in trying to place the blame. This disappearance of human responsibility is one of the most disturbing aspects of the whole thing. It's as if human beings had evaporated, and their places were taken by computers, and all day you and I have sat here, fighting not each other, but rather the big rebellious computerized system, struggling to keep it from blowing up the world."

NUCLEAR POWER IN POPULAR CULTURE

In the 1960s, a number of films began to exploit the impact of nuclear technology in areas other than war. In *The Day the Earth Caught Fire* (1961) atom bomb tests affect a shift in the earth's axis which sets the earth on a collision course with the sun. The event is shrouded in secrecy, government cover-ups, and denials ("there's nothing to be alarmed about"), until unusual climatic perils reveal the truth. Similarly *Crack in the World* (1965) has a scientist fire an atomic warhead into a fissure in the earth's crust in order to retrieve abundant energy from the center of the earth. Instead he splits the earth in two. This film begins to touch on the use of the atom to produce energy. But it was not until the mid-1970s when nuclear power was an important focus of public dispute that producers realized the potential for entertainment and sensationalism in the apparently routine and peaceful use of nuclear power.

Most early images of nuclear power came from popular science, and they contrast sharply with the apocalyptic visions of science fiction. Frederick Soddy himself had written in 1909 about the "energy of radium," describing it as a solution to the inevitable exhaustion of coal. With an optimism that was to characterize the early views of nuclear power, Soddy wrote of the atom as humanity's salvation: "A race which could transmute matter would have little need to earn its bread by the sweat of its brow . . . such a race could transform a desert continent, thaw the frozen poles, and make the whole world one smiling Garden of Eden" (Soddy, 1909). Drawing from Soddy, H. G. Wells's *The World Set Free* juxtaposed the devastation of atomic warfare with the hopes of nuclear-powered prosperity, suggesting the range of social consequences that could follow the development of atomic technology. Wells, however, was not sanguine about nuclear energy. He describes "the bright side of the new epoch in human history." But then "beneath that brightness was a gathering darkness, a deepening dismay. If there was a vast development of production, there was also a huge destruction of values . . . accumulated disaster, social catastrophe" (Wells, 1914).

Wells's caution, however, was not typical, for after the Second World War the optimism about the benefit of nuclear power grew into a fervor. The atom had won the war; it would now keep the peace. It was a magical source of cheap energy, a panacea for the world's problems. A genre of so-called inadvertent science fiction developed, as popularizers who often were themselves scientists predicted that automobiles and heating systems would be run by atomic piles and that poverty would end in the future world of plenty. The idea of social equity grew with the imagery of atomic innovation (del Sesto, 1980). The science editor of the Scripps Howard newspaper chain described how chunks of uranium mounted on towers would create artificial suns controlling the weather. "No baseball game will be called off on account of rain in the Era of Atomic Energy. No airplane will bypass an airport because of fog. No city will experience a winter traffic jam because of snow. . . . Mankind will move into an era as different from the present as the present is from Ancient Egypt" (Perkins, 1977).

Subtly the word "atomic power" changed to "nuclear power." Dr. Harold Fisher (154: 21), President of the American Chemical Society, wrote of nuclear-powered automobiles, airplanes, washing machines, and even wrist radios. Science writers in *Popular Mechanics* worked out the details of nuclear-powered engines even while publishing do-it-yourself designs for bombshelters. Nuclear power was expected to "end the world's hunger," turning deserts into agricultural regions as atomic engines were installed where no other power sources were available. They would permit the development of new regions and end inequality throughout the world. Nuclear energy, according to some writers, would even allow crops to be grown underground. Science writer William Laurence (1966: 41), who had observed the first bomb test, promised a land of milk and honey as a result of nuclear power. Again the new technology was envisioned as a panacea to social ills.

The popularizers were not fiction writers, and their point was not to entertain. Yet their product was also divorced from reality, albeit at the opposite extreme. However, in the 1970s, as public controversy about nuclear power widened, the extremes began to converge. Critical novels appeared centering on these themes and were written in documentary style that blurred the distinctions between fantasy and the real world. Docudrama turned fictional events into plausible scenarios. Thomas Scortia and Frank Robinson wrote *The Prometheus Crisis*, a disaster thriller describing a mysterious reactor meltdown in a Los Angeles nuclear power plant. The preface to the novel urged that nuclear power be stopped. Paramount Films bought the film rights in 1975, but under pressure from the nuclear industry never produced it and sold it to an independent producer (Smith, 1977: 20). A 1974 novel, *NUKE*, is a melodrama about a super breeder reactor with an ever-expanding production of energy that has to be used or else the breeder will explode. The government agency running the show is called the Bureau of Atomic Development (BAD). John Fuller wrote a fictionalized account of the 1966 breakdown of the Fermi reactor called *We Almost Lost Detroit* (1975). It describes the enormous difficulties of coping with accidents. *Nuclear Catastrophe* (1977) traces the lives of the people who are victims of an explosion at a California plant caused by a crack in the reactor pressure vessel. These works pick up on many of the political and organizational questions that were being raised by the anti-nuclear movement such as the problems of implementing regulations and the lack of emergency evacuation plans. In *Nuclear Catastrophe*, the utility manager is the first victim of the disaster, suggesting that the problem lies not in the evils of the establishment, but in its blind faith in the infallibility of nuclear technology.

The 1979 film, *China Syndrome*, is a fictionalized quasi-documentary account of an accident that turned out to be remarkably similar to the subsequent Three Mile Island accident. It conveyed the political issues of concern to the anti-nuclear movement, and it dramatically highlighted the pressure on the utility to restore the reactors to operative status. It also captured the secrecy, the inadequacy of quality control, and the cooptation of the opposition to nuclear power. Overriding all these themes was once again the persistent image of a technology running out of control. *China Syndrome* managed at once to be an entertainment thriller, a work of science fiction, and a political advocacy tract. Unlike the earlier nuclear films, it grossed millions of dollars. Subsequently the producers of *The Prometheus Crisis* obtained funding, and a fictionalized documentary of the Karen Silkwood case has been produced. A spy thriller about a formula that will turn radiation waste into plutonium is in production.

By the late 1970s television, too, began to fictionalize the atom. The "Lou Grant Show" included an episode involving a nuclear company employee who was murdered while compiling evidence of safety violations. Another episode depicted a terrorist group threatening to detonate a home-made plutonium bomb. After Three Mile Island, television programs on the atom proliferated. Old films were revived, and the soaps depicted a variety of nuclear thrillers, many of which turned on questions of reactor safety. In 1979 in the three months between April 7 and July 7 no less than fifty-five television shows used the atom as a dramatic device.

WE LIVE ON IMAGES

The atom in popular culture appears as a symbol of catastrophic disaster representing the mysterious power of transmutation and our vulnerability to extinction. Above all, the atom appears as technology out of control. These are powerful images, and the anti-nuclear discourse exploits them effectively to protest against nuclear power. Their importance in the anti-nuclear discoruse should not be underestimated. They play on an ambivalence toward technology that is deeply rooted in American culture, "a culture," according to literary scholar Leo Marx, in which technology "is at once a progressive force enhancing civilization and a moral force . . . a dark, fateful, well-nigh apocalyptic idea" (1980: 48–49). But the atom also evokes special associations of its own. Historian Spencer Weart suggests that the deep anxiety surrounding nuclear power today is a modern manifestation of deeply rooted fears that have long been associated with transmutation and alchemy. "There is abundant evidence that the symbol of transmutation calls to mind the great theme of death and resurrection . . . the moment one brings up the idea of transmutation, one must deal with ancient transformation and shattering dangers right up to the end of the world" (Weart, 1980).

Psychoanalyst Robert Lifton sees the atom as a symbol of the "death of the species," representing "an imagery of extinction," a form of "self-extermination with our own technology." Much of human history, argues Lifton, can be understood as "the struggle to achieve, maintain and reaffirm a collective sense of immortality under constantly changing psychic and material conditions." Thus, in raising doubts about our continuity as a species, the atom even in its manifestation as a source of energy represents a fundamental and ultimate threat (Lifton, 1980: ch. 22).

Why have these images that are associated for the most part with war assumed such political salience in the debate over nuclear power, with the peaceful use of the atom? Why are nuclear power plants as well as weapons endowed with these apocalyptic metaphors of war? Obviously public concerns about nuclear power were not directly caused by horror films. Rather, they followed the visible increase in the number of nuclear plants scattered throughout the country. Then, specific events like the Three Mile Island accident and disputes among scientists about reactor safety helped to trigger public doubts. While military policy is often protected from criticism by counterbalancing concerns about national defense, nuclear plants are a visible, proximate, accessible target. The image of the atom, though derived from war, is available and easily transferred to this concrete manifestation of the atom today.

Popular films and fiction have tapped fundamental anxieties, feeding unpalatable images almost intravenously to the passive viewer. Nourishing the imagination, they have built a level of psychic contamination around the image of nuclear power, and they have thus helped to create the symbolic environment in which the anti-nuclear movement has gained significant political force.

11

Counterpoint: Reactions to the Atom: Mutants versus Meltdowns

ALLAN OLMSTED

I was chosen to comment on Professor Nelkin's paper because I teach mass communications, although I seldom go to movies. Also, as my long-suffering colleagues are aware, I spent considerable amounts of time recently listening to some 375 intervenors speak to an environmental assessment panel that was reviewing the siting of a proposed uranium dioxide manufacturing plant in Warman in southern Saskatchewan. One additional qualification is that I read widely in the areas of science fiction and fantasy. This last led me to ask the organizers if I could open my comments today by stating "I am a giant among Hobbits." While they said I could not, this does indicate a personal bias relevant to the following discussion. By preference, I fear small, sneaky, invisible things and do not normally think on Professor Nelkin's apocalyptic scale. Monsters exist in my world, but on the order of the Love Canal, not mutated iguanas.

Professionally, I was trained in a very positivistic school of sociology, which among other things, insists that data gathering should proceed in a manner that allows collection of data which may disprove, as well as prove, a hypothesis. Therefore, I have disliked the combination of armchair science, social problems, and popular culture which led to the discovery that songs such as "Puff, The Magic Dragon," "We All Live in a Yellow Submarine," and "Lucy in the Sky with Diamonds" were probably causally linked to drug addition among pop music listeners. Similarly I disagree with the basic thesis contained in Professor Nelkin's paper.

Before discussing what I take to be some contradictions and alternative explanations in Nelkin's work, let me outline some points of agreement. First, there certainly is a persistent, growing mass movement which can generally be defined as anti-nuclear. Nuclear power is probably the only issue sufficiently salient to have created and maintained such a movement. I can also accept the "apocalyptic image" of nuclear power, if we keep in mind that apocalyptic images are those "forecasting or predicting the ultimate destiny of the world in the shape of future events." Whether popular culture is predicting ultimate freedom and power, or ultimate destruction, Professor Nelkin's selection of authors and titles demonstrates the perceived potential of the atom as entertainment.

The catharsis involved in monsters and catastrophes, whose "extraordinary level of fantasy," according to Professor Nelkin, "allowed the observers to dissociate themselves from the events," is an acceptable premise. Yet this must be qualified. First, we must remember that monsters have been a form of popular entertainment preceding books and movies. Second, they have continued since the atom has become serious. *Jaws*, *Claws*, *The Towering Inferno*, and a recent made-for-television movie called *Alligator* indicate that this genre is alive and well. Professor Nelkin's sampling overlooks this, while seeming to suggest that the atom has lost its cathartic ability to provide fantastic entertainment. Why has the atom become ordinary, mundane, and fearfully real? Another area of agreement is that in the latter days of the atom as entertainment, it is used to represent technology out of control. Certainly *Dr. Strangelove* or *Fail Safe* would be less dramatic if the technology of the automobile or coal mining were used as symbols.

Professor Nelkin's conclusion is that "popular films and fiction have tapped fundamental anxieties, feeding unpalatable images almost intravenously to the passive viewer." As a result, nuclear power plants are associated, via "psychic contamination," with the atom as a symbol of catastrophic war. Protests grow around nuclear generators, as the military is protected by concerns over national defense. A symbolic transfer is achieved, allowing passive viewers of Grade B movies to actively protest against the "atom," thus reducing subconscious anxieties.

Throughout this paper I have had the feeling that the word "apocalyptic" has been associated with the holocaust and other large concerns. If we accept the definition of predicting the destiny of the world, and so on, cited above, and link this with the North American ambivalence toward technology cited by Professor Nelkin, it would appear that nuclear technology is not the only technology that is out of control and apocalyptic. Coal power, water power, herbicides and pesticides, and the family automobile have also raised considerable fear in the general public. Is the concern with pollution, highway safety, and fuel supplies associated with subconscious fears caused by "My Mother the Car" and the "Dukes of Hazzard?" I suggest that we could make this claim as validly as Professor Nelkin makes the association between popular images of the atom and the mobilization of the anti-nuclear movement. Surely to create apocalyptic images of DDT and Agent Orange, we need cite only one "popular" science book—Rachel Carson's *Silent Spring*.

Professor Nelkin's research does not account for the fact that many individuals still perceive nuclear power as safe and efficient (25 percent in the public hearing in Saskatchewan mentioned earlier). Do these people not go to movies? Or does familiarity through working in the nuclear industry protect these individuals from intravenous information via popular entertainment?

Finally, Professor Nelkin dismisses "chemical plants, liquid natural gas storage facilities and toxic waste disposal sites" as capable of arousing local concerns but not of arousing apocalyptic images. With most major rivers and lakes in Canada already contaminated by these activities, a series of local concerns may link together for a much larger response. Radioactive waste lasts longer, is indeed associated with fears of mutation, but, except for creating movie monsters, water pollution has very low and slow production values as a movie gimmick. Nevertheless, it does arouse real concern among large numbers of people.

Let me briefly suggest some characteristics of the masses and the media which suggest an alternative model of the growth and persistence of the anti-nuclear movement.

Our colleague, Otto Larson, has suggested that the great increase in mass communication, particularly television, has greatly widened the public to be taken into account with respect to any public issue. In addition to entertainment, Harold Lasswell (1948) stated that the mass media are involved in surveillance of the environment, correlation of the parts of society in responding to the environment, and transmission of the social heritage from one generation to the next. A truism in the media is that bad news is generally a more attractive news item than good news.

With respect to the audience or public, there is increasing evidence that it is not formed of a large number of isolated individuals (a mass) but rather of individuals who are members of social groups. These groups are predominant in forming individual attitudes and values. For example, at the Warman refinery hearings, intervention came from virtually all religious groups in the area. (Many individuals intervened on the basis of their religious affiliation.) These groups were joined by environmental, alternate energy, and political groups.

Such shared attitudes and values have been shown to influence the relationship between audience and medium in the following ways. Individuals will consume information and use the media most consistent with their existing attitudes. This is known as *selective exposure*. When the selected media presents information contradictory to existing beliefs, members of the audience shape the information to agree with their beliefs. Anti-nuclear audiences see Three Mile Island as a disastrous incident. Pro-nuclear groups point out that the fail safe mechanisms worked, and there was no disaster. This is called *selective perception*. Finally, those pieces of information consistent with beliefs are remembered. Anti-pollution groups remember all the spills, pro-industry groups remember all the hours of safe operation. This is called *selective retention*.

Two final thoughts about the social base for the anti-nuclear movement are necessary. One human way to gain power is to form coalitions. Often these are formed on the basis of "the enemy of my enemy is my friend." Second, social movements exist in a larger social context. We are operating in a time of anti-war and anti-pollution movements. Others seek to improve civil rights and to improve the condition of the Third World and native peoples economically, as well as politically. Some groups represent alternate future lifestyles, although in a less than apocalyptic way. For each of these movements, the nuclear cycle is of considerable relevance and produces coalitions against the nuclear cycle.

In a week of media surveillance, it is not unlikely that one will hear about, read about, and/or see pictures of wars and anti-war protests. Incidents of pollution are reported from mines, refineries, waste disposal sights, and transportation links. Civil rights are reportedly violated, and Third World countries are subjected to war, famine, disease, and violent acts of nature. Native peoples lose land rights and are offered dangerous and low-paying jobs due to industrial development, in addition to the earlier problems. Hopes of a breakthrough in organic farming and alternate energy are occasionally reported but have low news value.

Media specialists and editors cannot, in the time and space allotted for integration, be expected to interpret all these issues for the audience. Often, this function is passed to experts. However, because experts are human too, their opinions differ. To compensate for this, North American and some other mass media allow equal time for opposing views. Some experts approve, others disapprove. Thus, the audience, and members of social movements, are free to reinforce their belief that they were correct all along.

Having listened to several weeks of testimony from both sides of the nuclear debate and having become extremely sensitized to media reports on this debate, I still cannot decide whether the nuclear debate differs in degree or in kind from cognate issues. I will close by giving some of the arguments presented by each side of the issue. These are reinforced by the non-entertainment functions of the media, and they create the coalitions that broaden, and strengthen, the anti-nuclear movement.

Nuclear weaponry and nuclear war are considered bad by everyone! They will kill everyone, pollute the environment, violate civil rights, let the survivors starve, alienate aboriginal rights, and ruin the chances of developing alternate futures. However, many people and experts believe that war is even worse if the other side has all the nuclear weapons, and that tactical nuclear weapons can limit nuclear war. Canadians avoid this argument by assuring everyone that Canadian uranium and Canadian reactors cannot, and will not (due to limiting nuclear weapon proliferation treaties), be used in nuclear weapons. This argument is countered by experts and investigative reporters who tell us that India and Pakistan have developed nuclear weapons from the by-products of CANDU reactors. The audience picks its argument. Any individual who believes that war is bad and dangerous can join the anti-nuclear war movement. Similarly anyone who fears that war is sometimes necessary may argue to continue nuclear weaponry.

Everyone agrees that all toxic pollution is bad. Some argue that radioactive pollution, at low levels, is the worst of all. It is harder to contain, lasts longer, causes more cancer, and may cause long-term genetic defects. Opponents of this view argue that technology will be found to contain emissions and wastes, and that low-level radiation is no more dangerous, indeed less dangerous, than acid rain, PCBs, and other existing pollutants. Once again, the mass media bring us a full array of experts, and we can all agree with some of them.

As all these familiar arguments were presented to me, I heard Mennonites, campus radicals, scientists, politicians, native persons, housewives, children, whole earth farmers, agribusiness farmers, business people, experts, and members of ethnic minorities quote mass media sources to prove that the only good atom is an unsplit atom. Any stage of the nuclear cycle can only increase the dangers we are currently facing.

During the same hearings I heard Hutterites, academic scientists, industrial engineers, politicians, farmers, business people, and experts quote media-transmitted sources to prove that the nuclear industry provides progress, clean and efficient energy, and that they also were against war and pollution. (Hutterites, the only group that claims it abstains from media use, thus support Professor Nelkin and reject my argument.)

Because nuclear technology (except weapons) is not yet in place, one side seemed to say, "let's not take a chance on another dangerous one"; the other side said that "with what we have learned from previous energy sources, we will not make the same mistakes." Because so much nuclear usage is potential, we have an issue that is alive and continuing, with plenty of ammunition for every argument. The issue at these hearings was a uranium refinery, not a generator or a missile site. However, the latter were considered as equal parts of the nuclear cycle. Perhaps if the issue had been a generator or a missile site, the fears would have reflected the catastrophic images presented by Professor Nelkin. However, the media operated as expected, and the audience followed the classic rules of opinion formation. Other technologies reinforced the fears of cancer and mutation and apocalyptic change. A global concern over incident after incident seemed to say that the apocalypse would be revealed through a series of whimpers all over the real world, rather than a bang from Hollywood.

Fictional monsters are still a source of escapist entertainment, but cracks, leaks, smogs, spills, incidents, and accidents are in all of our local neighborhoods and communities. They threaten our morals, our food production, our reproductive capacity, our communities, our lives. Nuclear technology is the latest, if not the greatest, technology to be produced for us by our society, which promised to free us and is now threatening us. A wide public, a series of local concerns, and disparate groups are using the mass media not just for popular entertainment, but to create and maintain a movement concerned with controlling the potential dangers of the atom, even at the cost of losing its potential benefits.

V

Public Policy, Determinism, and Change

All three major papers in this section deal at length with public policy as it relates to technological change and environmental impact, especially in the context of employment. William Leiss examines the coming "information revolution" that is alleged to be determined by discoveries in computer technology. As Professor Leiss notes, governments in Canada, the United States, and France have commissioned studies to underwrite the social response said to be dictated by these innovations. Professor Leiss challenges the idea that such changes are either desirable or inevitable. He would have us resist the urge to see public policy solely as an ameliorative response to trends that must be accommodated in some fashion. In his response, William Reeves expands on the initial paper by exploring the hypothesis of technological determinism as it applies to the manufacturing economies. He points out that the response to new technologies differs in peripheral entrepreneurial industries versus central integrated industries. In neither context is technological change inevitable. Rather, technological innovations are catalysts that are more liable to create effects under conditions of competition.

In his treatment of the microelectronic revolution Stephen Peitchinis also examines the impact of the new technology on employment. He focuses on the need to devise an appropriate social policy to ameliorate the labor force dislocation following the widespread introduction of microelectronic processors. Specifically Professor Peitchinis suggests that a reconception of the nature of work to cover voluntary social service might be underwritten by the new wealth created by the computer revolution.

A rejoinder by Donald L. Mills qualifies the concept of "jobless growth" that is the premise of Professor Peitchinis's paper. Professor Mills suggests that under "jobless growth" the growth goes to the wealthy and the joblessness goes to the socially marginal elements of society. Professor Mills cites evidence that the early social casualties of microelectronic processors are women, the young, and other disadvantaged groups. As to the policy response suggested by Peitchinis, Mills views this as rather naive and idealistic. He suggests that change in basic social values concerning stratification and reward is unlikely to come about as simply as Peitchinis suggests. Such change will likely be resisted by many parties in society. It is difficult for Mills to accept the scenario of remuneration for what has always been defined as non-work. He ends by reminding us that far-reaching social structural changes are required if we are to experience the rather utopian future described by Peitchinis. Professor Mills's position reiterates the concern for equity found in earlier comments by Allan Schnaiberg, W. G. Carson, Phil Elder, and Joseph DiSanto.

In the final paper of this section, Professor Schnaiberg returns to questions of equity and employment in his analysis of A-T. Where Leiss responded to the presumed inevitability of the information society, Schnaiberg's point of departure is the presumed feasibility of an A-T society. A move toward A-T, however humane, reduces the productivity of industry and undermines the redistributive component of the earlier environmental movement. This follows from elimination of the economies of scale found in the traditional treadmill of production. Professor Schnaiberg suggests that a viable A-T movement could exist only on the periphery of the treadmill, subsidized by the surplus of the treadmill. He suggests that it is doubtful that an A-T model would be adopted either eagerly or successfully by all workers in all circumstances, and he reminds us that workers are not all angels nor managers demigods. It is in his insistence on such Realpolitik that Schnaiberg's paper takes up Mills's earlier comments. Schnaiberg, too, grounds his critical remarks in a social structural analysis of real-world experiences of employment as opposed to the more hortatory and visionary discussions of E. F. Schumacher or Amory Lovins.

Finally, in his response, physicist Harvey Buckmaster speaks both to the means, as well as the relations, of production under A-T. Professor Buckmaster attests to the empirical and practical soundness of A-T designs especially in the Third World. While acknowledging the difficulties entailed in changing social relations, Professor Buckmaster reminds us that the utopian ideals of the nineteenth-century Fabian Movement, however seemingly unrealistic, nonetheless were instrumental in the formation of the modern labor movement. Consequently he recommends that, despite its shortcomings, the A-T movement not be dismissed as a force of social change. Identification of its weaknesses should lead us to attempt to overcome them. It should not lead us to give up the attempt to reform our society and step off the treadmill.

12

Under Technology's Thumb: Public Policy and the Emergence of the Information Society

WILLIAM LEISS

The notion that technological innovation can produce qualitative changes in social relations is a hallmark of modern thought or, to use the current jargon, of "modernity." Its great early propagator was Francis Bacon, who regarded conventional politics as a zero-sum game where benefits could be extracted by one party only at another's expense; the conquest of nature through science and technology, however, promised to overcome this limitation and to deliver an ever-increasing supply of unqualified benefits.

For Bacon the essential promise of technology was not that it could address directly the traditional sources of human unhappiness, but that it could gradually render them irrelevant. Since in any case a significant proportion of the problem was owing merely to intellectual confusion, what today we are wont to call "ideology," those who remained mired in such confusion would appear more and more ludicrous in relation to the tangible increases in human welfare won by the new technology. Two and one-half centuries later, the very title of a 1966 journal article by an eminent political scientist, Robert E. Lane—"The Decline of Politics and Ideology in a Knowledgeable Society"—shows how much vitality this mode of thinking still possesses (Lane, 1966).

In the second phase of its development, beginning in the nineteenth century, a new variation emerged which emphasized the inevitability or necessity of social changes induced by technological innovation. Employing a crude version of the evolutionary metaphor, this variation insisted that societies must "adapt" to their new "environments" or suffer the consequences. This reinforced the fundamental thrust of technocratic thinking, which favors the displacement of questions about value choices, regarded as ultimately ideological, by a single overriding determinant for decision-making that is "objective" and quantitative in nature: efficiency in the allocation of resources. Since what is happening is allegedly inevitable, whether or not we like it is irrelevant. The 1981 Department of Communications report, *The Information Revolution and Its Implications for Canada*, maintains that "like the industrial revolution, the information revolution is unavoidable. Consequently, the objectives of public policy should be not to prevent the revolution from occurring, but rather to turn it to our advantage" (Serafini and Andrieu, 1981:13).

The mention of public policy in this context signals the advent of a decisive new phase in technocratic thinking. In earlier phases the inevitability or necessity allegedly inherent in technological innovation was presumed to have a direct, unmediated impact on social relations. Of course the economy, where the determination of the efficient allocation of resources occurred, was the actual transmitter for these impacts. In the current phase *public policy is supposed to serve historical inevitability by facilitating a favorable social response to it.*

Two quite different sets of circumstances have produced the new phase. First, the rate of innovation and turnover apparently has accelerated; this, together with the intensity of international competition, means that societies must respond much more quickly than they did in the past; and public policy must provide some "grease" to insure a faster response time. Second, at least some interest groups are much better able to defend themselves by protecting their income, status, and influence against innovations that threaten to erode their relative advantages than they used to be. The enhanced ability of social groups to articulate their interests, and to require governments to protect those interests to some extent, places an important intervening variable between technological innovation operating through the economy and social relations. Public policy must seek in part to persuade us to acquiesce in what we can no longer be forced to accept, at least not without a protracted struggle. And that is just the point: because, even if, as believed, the innovating forces must triumph in the end, they will have achieved only a pyrrhic victory, for the economic advantages of early entry will have been lost, and our society will drift further and further away from the "action" as each successive wave of innovation rolls in.

The concepts of "information revolution," "information economy," and "information society" constitute an important new stage in the tradition of technocratic thinking in modern society. In large part their importance lies precisely in how perfectly they represent this tradition. They enable us to see clearly what role public policy is thought to have in the interaction between technology and society: namely, to "soften up" public opinion so that a compliant social response to a new technology may be delivered. It seems to matter little to its propagators that their show includes some quite outrageous sleight-of-hand routines, for what is important is not the act's constitution but its effect: in this case, as in technocratic thinking generally, to attempt to persuade us that we are free to choose only the *timing* of our submission.

The effort at persuasion introduces a nice circularity into the process. If we can be cajoled into believing that some future state is inevitable, and to alter our behavior in order to conform with its anticipated requirements, the end result will be a retrospective proof of the prediction's accuracy. These are its principal steps:

1. *Analysis* develops a conceptual model, namely, the concept of the "information society" whose objective is to influence
2. *Policy* initiatives that will create favorable conditions for shaping a
3. *Social Response* that over time results in changed social behavior and new
4. *Behavior Patterns* that resemble those originally predicated as desirable in the
5. *Analysis* itself, thus confirming the model's predictions about what was "inevitable."

At the time when many thought that it was salutary for society to be utterly at the mercy of the marketplace's allocative mechanisms, and that "interface" by public authority was to be avoided, the apparent "necessity" in the process required no further justification. Public policy today, however, as the explicit voice of public authority, abdicates its responsibility and loses its *raison d'être* when it limits itself to the "recognition of necessity." For there is no necessity, strictly speaking, in social events; rather, they represent the outcomes of individual and collective choices (including both conscious and unconscious motivations) that rest ultimately on fundamental values. From this standpoint it is the duty of public policy discussions to clarify the full range of choices to be made, and their possible impacts on values, so that enlightened decisions may be made about the future directions of social change.

The "information society" and its associated notions will be examined in this paper as essentially the most recent achievements in the long tradition of technocratic thinking. They evolved in a three-stage process during the past twenty years or so: from the "technological society" to the "knowledge society" and/or "service society" to the "information society." The chief distinction between the information society and its predecessors is that, in the attempt to achieve an anticipatory public policy commitment through "co-ordinated programs to increase public awareness of the benefits of the new information technology," as the Department of Communications report (1981:96) puts it, there is an explicit strategy to steer social responses to technological change in a comprehensive way. This development demands careful analysis and reflection.

This examination of the information society will be made in the context of a general standpoint on technocratic thinking to which I will return in the concluding section. It has three major features:

1. Social policy options in contemporary advanced industrialized societies generally are constrained by zero-sum solutions, as Lester Thurow (1980) has argued in *The Zero-Sum Society*. This constitutes a fundamental obstacle to the basic thrust of technocratic thinking, namely, that technical solutions can displace "political" and ideological disputes.
2. Even if it is true, to some extent, that powerful pressures are exerted on industrial societies to adopt new technologies, given their integration into an international economic structure, these societies can and should resist the notion that they must adapt to the new environment in any predetermined way. The mode of adaptation for societies is not fixed in advance but can be made responsive to reflective processes through which societies retain a measure of freedom from necessity and of choices based on the autonomy of value systems.
3. Public policy documents on the interplay of technology and society should embody a strategy of "social accounting" that eschews a one-sided statement of the anticipated benefits of new technologies only. Rather, this accounting should encompass a frank assessment of what is expected to be lost as well

as what is to be gained, which would enable us to attempt to preserve by some other means those things of enduring value to us.

BACKGROUND

In the 1960s there emerged an extensive set of publications and organized research projects on the theme "technology and society." The most prominent project and its associated publications was the Harvard University Program on Technology and Society (about 1964–71), funded by a large IBM grant and headed by E. G. Mesthene, who summarized its overall outlook in his book *Technological Change* (1970); the best known single work was John Kenneth Galbraith's *The New Industrial State* (1967). Much of the literature was a response, in one form or another, to the exasperating and formidable tract by Jacques Ellul, *The Technological Society* (1967), originally *La Technique* (1954); in general the writing brought to fruition a theme that had been articulated shortly after the end of the Second World War, as represented best in a UNESCO symposium (still worth reading today) published in *The International Social Science Bulletin* in 1952.

The key concept in this literature is that of "technological society" itself. It was intended to suggest that a new "type" of society had emerged: one that was distinguished by the centrality of continued technical innovation in its midst, one that was qualitatively different from all earlier societies by virtue of the centrality of technology, and thus one that required of its social relations that they be capable of responding continuously and, of course, favorably to technology's relentless upwelling. Mesthene (1967:59) gave these propositions their most extreme formulation: "Technology, in short, has come of age, not merely as a technical capability, but as a social phenomenon. . . . We are recognizing that our technical prowess literally bursts with the promise of new freedom, enhanced human dignity, and unfettered aspiration."

Many writers emphasized the key social role of "organized knowledge" in the technological society, and this quickly became an important subsidiary theme in the literature. Daniel Bell developed this angle most fully, and his *The Coming of Post-Industrial Society* (1976:xii) is the major transitional work in the tradition under examination here. Bell's notion of post-industrial society laid the foundations for the later concept of the information society: "Broadly speaking, if industrial society is based on machine technology, post-industrial soceity is shaped by an intellectual technology. And if capital and labor are the major structural features of industrial society, information and knowledge are those of the post-industrial society." Bell's chief emphasis, however, was on the transition from a stage of modern society dominated by goods production to one in which "services" were fast becoming the dominant economic sector. A few years later others reconceptualized the process, and the "service society" was transformed into the information society with but a minimum of further tinkering.

The broad background theme in this literature is composed of the so-called imperatives of technology. Industrial society developed by reorganizing the matrix of social relations so that there would be a receptiveness to continued technological innovations; it did so largely by increasing the flexibility and responsiveness to market conditions of the factors of production. Originally, therefore, modern technology was the means for enhancing productivity, which in turn was the means for bettering the general welfare. In ironic reversal, however, the means became autonomous and thus increasingly an end-in-itself: technology imposed the hegemony of its own supreme value (that is, efficiency) on society generally. Ellul's famous book tracks this process throughout all the crevices of social life and bitterly laments the outcome. Most other commentators are more phlegmatic. Galbraith (1967), for example, argues simply that, once a commitment to a high level of industrialization is made, the "imperatives of organization, technology and planning" overawe ideologically grounded differences (that is, capitalism versus socialism) and give rise to a basically identical outcome dictated by the nature of the technically oriented infrastructure itself.

For most writers this outcome is not the unmitigated disaster lamented by Ellul. Precisely the opposite: technology opens vast new "possibilities" for mankind, and we are free to turn these opportunities to our advantage in any ways we choose. What we must do is simply adapt ourselves to the new technologically formed environment, to develop a "readiness" to exploit the myriad new opportunities for action. The well-known author Donald Schon (1967:xiii) stated this point well: "But, as we are learning, technological innovation belongs to us less than we belong to it. It has demands and effects of its own on the nature and structure of corporations, industry, government-industry relations and the values and norms that make up our idea of ourselves and of progress."

Since the notion of the information society is a legitimate offspring of this tradition, its distinguishing features are those it merited.

Qualitative Change. There is a major shift in socioeconomic structure, defined by broad changes in occupations and in the social significance of productive factors. Stated most provocatively, the idea is that knowledge replaces capital as the governing factor. This precipitates equally significant changes in decision-making processes, including control over the political process and how political power is exercised.

Social Responses. Society is encouraged to respond positively to "new opportunities" that can and should have wide-ranging impact on lifestyles and quality of life. In Canada, recommendations are made for pubic policy initiatives designed to encourage the desired social responses.

THE INFORMATION SOCIETY

In 1978–79 GAMMA, a university-based research group in Montreal, undertook an "Information Society Project" funded by the federal government. In the integrating report prepared by the group's director, Kimon Valaskakis, the following definition is offered: "An Information Society is a set of social relationships based on an Information Economy. In turn, the Information Economy exists whenever over 50% of the Gross National product belongs within the broad information sector" (Valaskakis, 1979:40). Also in 1979, Gordon B. Thompson of Bell-Northern Research prepared a report for the Institute for Research on Public Policy entitled "Memo from Mercury: Information Technology *is* Different." And, as previously mentioned, in 1981 the federal Department of Communications issued a report written by S. Serafini and M. Andrieu, *The Information Revolution and Its Implications for Canada*" (Serafini and Andrieu, 1981).

Daniel Bell's *The Coming of Post-Industrial Society* is the major undertaking in comprehensive social analysis that paved the way for the concept of the information society. Bell relied heavily on some statistical evidence (from OECD (Organization for Economic Cooperation and Development) and other studies) to argue the growing preponderance of scientific and technical knowledge in the economy and society. Subsequently there appeared an elaborate study, *The Information Economy* by Marc Uri Porat (9 volumes, 1977), on which the reports listed above have rested their case for the existence of the information economy. "The information economy" is the mainstay for the ensemble of social impacts that are said to make up the information society.

The Post-Industrial Society

Bell (1973:14) described five distinguishing characteristics for post-industrial society:[1]

1. Economic Sector: The change from a goods-producing to a service economy;
2. Occupational Distribution: the pre-eminence of the professional and technical class;
3. Axial Principle: the centrality of theoretical knowledge as the source of innovation and of policy formation for the society;
4. Future Orientation: the control of technology and technological assessment;
5. Decision-making: the creation of a new "intellectual technology."

Let us examine these claims, especially the first two.

The traditional classifications in the U.S. Department of Labor statistics used by Bell and the percentage of the workforce in each category are as follows (rounded):

	1900	1960	1974
White collar workers	17.5	42.0	48.5
Manual workers	36.0	37.5	35.0
Service workers	9.0	12.5	13.0
Farm workers	37.5	8.0	2.0

When these categories are reanalyzed into just two sectors, goods-producing and service-producing, the figures for 1968 in the United States are: Goods, 36%; Services, 64%. In this far more simplified stratification, the actual composition of the two, especially the service-producing, is crucial. The service sector includes the following: transportation and utilities, 5.5%; trade (wholesale and retail), 20.5%; finance, insurance, and real estate, 4.5%; services, 18.5%; government, 14.5%. Services are composed of personal (laundries, garages, hairdressing, and the like), professional (lawyers, doctors, and accountants), and business (office equipment, cleaning, and so on). The goods-producing sector includes only those directly employed in mining, construction, manufacturing, agriculture, forestry, and fisheries.

The heterogeneous character of the service sector should give one pause at the outset. First, replacing the white collar-manual distinction with the service-goods one can be misleading if one does not remember that the two are not symmetrical (that is, white collar and service are not identical). Second, and more importantly, a large proportion of the service sector is a necessary and integrated part of goods-producing activities; this is especially true of transportation and utilities, but it applies to part of wholesale and retail trade, government, and business services as well. To some extent the inflated service sector reflects only the greater internal complexity of the goods-producing sector itself, which requires a higher level of infrastructural support now than it did earlier. In short, the goods-service distinction is a conceptual dichotomy, not a simple reflection of economic activity itself; one must be careful not to make too much of it.

The second point has to do with the "pre-eminence" of the professional and technical occupations in the economy as a whole. Included in the white collar category, their percentage figures increased from 4.3% in 1900 to 10.8% in 1960 to 14.4% in 1974. Who are they? For the United States in 1975, the breakdown for this category was as follows: scientific and engineering (including social scientists), 15%, of whom three-quarters were engineers; technicians (excluding medical and dental), 11%; medical and health professionals, 17%; teachers, 23%, of whom three-quarters were elementary and secondary teachers; and general, 34% (accountants, lawyers, media, architects, librarians, clergymen, social workers, and so on).

What conclusions can be drawn from the data? The more traditional categorization (white collar, manual, and so forth) shows in fact the remarkable stability of the "manual" sector as a percentage of the total labor force throughout the twentieth century. The principal redistributional shift has been the precipitous decline in farm workers, and these have been absorbed almost entirely into the white collar sector. Although the white collar sector is now numerically predominant, nothing in the numbers per se implies any kind of qualitative change in social influence. As Bell himself notes, traditional agrarian societies also have large service sectors, for example, household servants, made up of low-status occupations. The service sector today includes, as well as the high-status professions, a large proportion of low-status jobs in retail sales, clerical/typist positions, minor bureaucratic functions, and the like.

Bell says very little about this matter because it is point 2 that is the real heart of his analysis, namely, the idea that the professional and technical "elite" of the service sector is "pre-eminent" in our "new kind of society." He says: "The central occupational category in the society today is the professional and technical" (1973:137). Even if we limit ourselves to his own framework for analysis, however, it is difficult to see why this is supposed to be so. In the first place, the breakdown presented above shows clearly that this group is by no means a homogeneous social group which has, or potentially could have, a sense of self-consciousness as a group and thus a distinctive social interest. The category itself is composed of sharply divided strata in terms of self-identity and perceived status differentiation; the category lumps together the high-status professional "elites" (doctors, lawyers, accountants, engineers, architects) along with teachers, technicians, nurses, social workers, and so forth.

Second, each of these two basic divisions seems to have much more in common, in terms of income and social status, with those in other categories than they do with each other. For example, the professional elites have much in common with the stratum of corporate and government-sector managers and the more successful small proprietors, who are included in a separate occupational category. There is very little, if any, evidence (Bell presents none) to suggest, as his scheme does in moving from point 2 to points 3–5, that the professional technical group as a whole is increasingly in charge of directions for social change through policy formation, or even seeks to be in charge, based on what Bell calls their ability to use "intellectual technology" (a fancy name for systems analysis, organizational theory, and the like).

The analysis is a classic case of special pleading. Bell highlights the pre-eminence of the professional-technical group by pointing to the dramatic rate of increase in this "sector" relative to others. For the 1958–74 period there was a 77.5 percent increase, as opposed to a 55 percent increase for white collar workers as a whole. Nonetheless, the clerical sector grew by 65 percent in the same period! There are also enough possible anomalies in this particular period, including the deliberate channeling of resources into education, space research, and military expansion, to introduce a need for caution. In any case, the highlighting is achieved by segregating occupation groups like the professional and managerial which in reality have strong affinities in contemporary society.

The ultimate conclusion is as shaky as the analytical premises on which it is founded. For Bell the distinguishing characteristics of post-industrial society signal "the emergence of a new kind of society," and, as in all such occurrences, this one "brings into question the distributions of wealth, power, and status that are central to any society" (Bell, 1973:43). The reason is that in post-industrial society knowledge has emerged alongside property to constitute the two "axes of stratification" for social relations. One does not denigrate the important social function of knowledge today if one concludes, that on the evidence submitted, this contention appears to be most implausible.

THE INFORMATION ECONOMY

The major Canadian policy documents concede the point that the concept of information economy provides the bedrock for that of information society. All studies of the information economy are derived from the pathbreaking work by Porat, published by the U.S. Department of Commerce in 1977. This work concluded that 46 percent of the gross national product (GNP) could be classified as information activity, and 53 percent of all income as income earned by information workers. (The data go back mostly to 1967). The following basic definition was employed (Porat, 1977, V. 1:2): "Information is data that have been organized and communicated. The information activity includes all resources consumed in producing, processing and distributing information goods and services." It is composed of six subsidiary types of activity: (1) information generation or creation; (2) information capture, or the channeling of information; (3) information transformation; (4) information processing (at the receiving end); (5) information storage; and (6) information retrieval.

The principal working hypothesis is that information activity is not an independent sector in the traditional sense, such as manufacturing, but rather is something that cuts across all sectors. Thus, the information component of all types of economic activity must be segregated, and the results combined to provide an overall picture. Parenthetically, Porat concedes that a similar operation could be performed with regard to educational activity, for example, since most types of economic activity have an educational or learning component. This type of approach does not exclude others that employ similarly broad or "synthetic" concepts. This is an important concession; not surprisingly, it does not reappear in the policy literature. When this reshuffling is done, there are six sectors: a primary information sector, two secondary information sectors, and three non-information sectors.

1. Primary Information Sector: a private market sector, including all of the computer and telecommunications industry, finance and insurance, media, and private education, as well as varying percentages of other industries.
2. Secondary Information Sector: (a) public bureaucracy; (b) private bureaucracy; including all "information support" activities in large organizations.
3. Non-Information Sectors: (a) households; (b) private productive sector; and (c) public productive sector (for example, Crown corporations).

The first sector accounts for 25 percent of GNP, and the third for 21 percent, for a total of 46 percent (McNulty, 1981:13ff.).

The proportion of information workers in the total labor population is classified by means of a three-part typology. The first part consists of knowledge producers (engineers, lawyers, most specialist occupations) and knowledge distributors, primarily teachers; the second, of knowledge users, including in one sub-category managers, administrators, and bureaucrats, and in another, clerical and office staff; and the third, of the direct operators of information-processing machines. By an elaborate segregation of activities in industries, Porat arrives at the figure of 53 percent as the share of total income earned by information workers. An Appendix to the Department of Communications report offers an overview of the major classifications and occupational types taken from an OECD study, "Report on Economic Analysis of Information Activities":

I. Information Producers
   A. Scientific and Technical: natural scientists, engineers, social scientists.
   B. Market Specialists: brokers, buyers, insurance agents.
   C. Information Gatherers: surveyors, inspectors.
   D. Consultative Services: architects, planners, dieticians, accountants, lawyers, designers.
   E. Other: authors, composers.
II. Information Processors
   A. Administrative/Managerial: judges, office managers, administrators.
   B. Process Control: supervisors, foremen.
   C. Clerical: typists, clerks, bookkeepers, receptionists.

III. Information Distributors
  A. Educators: teachers (all levels).
  B. Communications Workers: journalists, announcers, directors, producers.
IV. Information Infrastructure
  A. Information machines: office machine operators, printers and associated trades, A/V equipment operators.
  B. Postal and telecommunications Workers.

Applying the OECD categories to the Canadian labor force for 1971, the Department of Communications report estimates that information workers accounted for 39% of total income earned as compared with 53% in the United States in 1967. The shares of each of the four major types of information activities were: producers, 19%; processors, 63%; distributors, 12%; and infrastructure workers, 6%. Unfortunately, the proportion of what is by far the largest type (processors, 63%) was not further disaggregated; since this involves two very different levels of occupational structure, namely, managers, including both administrators and clerical staff, a further breakdown would have been helpful.

Two general considerations arise immediately. First, the Department of Communications data show a lack of correlation between the size of the information sector in national economies, on the one hand, and overall economic performance on the other. The United States leads the way, in terms of size, followed by Canada and the United Kingdom; but West Germany is significantly lower, and Japan lowest of all. This is almost certainly related to productivity figures. During the past decade industrial productivity in the United States grew 90 percent, whereas office productivity grew only 4 percent; since so much of the information sector is office work, the Department of Communications report concedes that "low growth in productivity may well have contributed to an expansion in information employment" (Serafini and Andrieu, 1981:19). If this is part of an "information revolution," something seems amiss, for it sounds a good deal more like a counter-revolution.

Second, much of what was said above in a critical vein about the service sector applies here as well. The information economy is a mixed bag of occupations; whether we see any unifying ingredient in the mixture depends very much on our level of tolerance for terminological laxity. In any event they are the same cat, with only slightly different stripes. The Department of Communications report tells us that in Canada in 1971, "76% of information workers belonged to the service sector" (Serafini and Andrieu, 1981:15). I hasten to add that this is a perfectly legitimate enterprise for social analysts: very little under the sun is actually new, and fresh understanding often is derived from nothing more dramatic than rearranging familiar facts to fit another paradigm. Exercises such as Porat's, and speculative treatises such as Bell's, challenge other ways of thinking about the contemporary phase of industrial societies; even if one cannot accept them, one can appreciate the testing of one's own aproach.

It is a different matter altogether when others attempt to compose marching music out of these scattered notes. The GAMMA report (see Valaskakis, 1979:40) would have us believe that "it is reasonable to assume that the Information Society is now as inevitable for the OECD countries as puberty is for an adolescent." It further warns us to eschew reliance on either market forces or uncoordinated public policy, and urges the federal government to adopt a "concerted and comprehensive information policy." The Department of Communications report concurs:

> The social and economic impact of the information revolution could be as profound as that of the industrial revolution. Many industrialized countries recognize this fact, as well as the need for comprehensive approaches to policy in order to deal effectively with the widespread changes expected to result from pervasive application of these new technologies. (Serafini and Andrieu, 1981:8)

Just what is this bandwagon on which we should all leap? To answer this question, I will limit the discussion to what is contained in the GAMMA report.

THE INFORMATION SOCIETY SYNDROME

We must take for granted that society reflects the character of its economy; thus, the information society appears at the moment when at least 50 percent of GNP can be attributed to information activity. This is said to be the same as "the new division of the economy into high-throughput sector (processing energy and materials) and an information sector (processing symbols rather than goods, bits rather than BTUs)" (Valaskakis, 1979:40). What is immediately apparent is that this refinement is simply a direct appropriation of the goods-services dichotomy, except that the service sector has been renamed the information sector; and although the opposition between processing materials and processing symbols is somewhat rough and arbitrary, it is at least suggestive. However, this is followed up by the first in a series of astonishing embellishments, namely, "that because of the generality of the Porat-inspired definition, many pre-industrial or non-industrial economies may also be considered Information Societies. Thus, all religious-based societies (where religious activity is an important proportion of total activity) would qualify. Religion is, of course, by its very essence, an information activity" (Valaskakis, 1979:40).

One wonders what Porat would think of these liberties; such are the risks of publication, alas. We should recall that Porat's definition of information refers to data; is religious knowledge "data" of this sort? And is religious activity part of the economic life of these societies, except in a metaphorical or indirect sense? One hesitates to make too much of this flight of fancy. But it illustrates well what is required when one wishes to construct the need for a "concerted and comprehensive information policy" on the basis of little more than a preliminary reclassification of occupational categories.

The GAMMA report forecasts three different "scenarios" or options for our information society: the telematique, privatique, and rejection scenarios. Each in turn has a number of potential variants. The telematique scenario envisages a highly centralized computer-communications system, employing unified data banks with omnipresent access through terminals located everywhere in homes, offices, and factories. For example, "the telematique home will be centered around the television set," which will of course be a two-way or interactive mode. In factories, robotics and CAD/CAM processes will prevail; offices as such would probably diasppear, since these activities could just as well be carried out through home terminals and portable "electronic briefcases."

Of three possible variants of the telematique, the first is where the information content in the system is largely trivial, that is, degraded mass culture; in the second, the content is "totalitarian," that is, imposed on individuals by dominant institutions. And the third?

> This third variant of telematique is, by definition, the most desirable one. . . . Obviously, the information traveling along the electronic highway should be in the "public interest." Unfortunately, no one has a clear notion of what the public interest is or should be. We face here a real and, at the same time, inevitable dilemma. We cannot ignore public interest as a policy-guide because

if we did, there would be no criteria to assess public policy. At the same time, most hasty definitions of that elusive concept will in fact be promoting one of other special interests of particular groups. The task of public policy must be nevertheless to explicate that difficult idea and base itself on it, however monumental that task proves to be. (Valaskakis, 1979:48)

It would be uncharitable to regard this vision as vacuous. Fortunately the dilemma alluded to is equally insubstantial. The "privatique" scenario also assumes a ubiquitous computer-communications infrastructure, but with decentralized data banks. What is envisaged is highly individualized interactions or small-group networks reflecting specialized interests and needs. Finally, the rejection scenario is what its name implies, the spurning of advanced communications technologies in favor of personal interactions.

CONCLUSIONS

The notions of information society and information technology, at least in their present form, show the impoverished state of technocratic thinking.[2] Erected on flimsy conceptual foundations, composed of hastily recycled terminology, and motivated solely by the conviction that the show must go on, they offer us the old routine: a new technology demands a response from us that is appropriate to its essence and modes of action. Despite the fact that apparently only its negative aspects (the trivial or "totalitarian") are discernible, we are told that there must be a positive or desirable aspect, although it is impossible to say what it is. And despite the fact that the desirable course of action is a mystery, we already know that we will be unable to discover it through either the operation of market forces, through which consumer preferences might develop, or uncoordinated public policy responses. A "comprehensive" public policy is allegedly necessary, so as to construct the framework within which appropriate, that is, preprogrammed, social responses will emerge. Since little or no positive content can be specified at the moment, the actual message—the sole guideline for the comprehensive public policy—can be stated in stark simplicity: Whatever is happening is inevitable, and therefore we should prepare for it (whatever it is).

Few will deny that the marriage of computer and communications technologies will have a noticeable, and perhaps significant, impact on occupational structures, industrial and office productivity, employment opportunities, and everyday life. Equally few should be so incautious as to assert that we are in the throes of an "information revolution" that will rival the Industrial Revolution's impact, or that major qualitative changes in social relations will occur as a result. Nothing that has happened so far should cause reasonably attentive (but skeptical) observers to doubt the evidence of their senses, which is that whatever is happening indeed can be attended to by a judicious interplay of market forces and uncoordinated public policy initiatives.

This moderate perspective is grounded in a more general outlook concerning the nature of major public policy issues in contemporary society. This outlook in turn is based on the conviction that it is not new technologies, or the "new possibilities" for action embodied therein, which are or will govern the definition of those issues or our responses to them. They are principally what may be called "allocative" issues, and the solutions to them, such as they are, have zero-sum characteristics. Examples are income policy, federal versus regional or provincial powers, the relation between employment and social status, environmental protection, and redistribution and inequalities. The great task for public policy is to assist us in finding reasonably civilized ways of dealing with such issues. If we can do so, we will discover that managing the social impact of new technologies is by comparison mere child's play.

NOTES

1. Compare what follows with the brilliant commentary on Bell's work in Krishan Kumar's *Prophecy and Progress* (1978: ch. 6).

2. In general what has already been said applies as well to the latest public policy document in Canada: Science Council of Canada, *Planning Now for an Information Society* (1982).

13

Counterpoint: Technological Innovation and Social Change

WILLIAM REEVES

I feel that it would be useful to summarize the basic arguments advanced by the advocates of the coming information society. The significance of Leiss's points becomes more obvious in the context of this review. I will conclude my discussion by elaborating on the ideas of market competition, technological innovation, and the notion of technological determinism.

Advocates of the information society have noted that the capital goods industries for data processing and telecommunications have rekindled competitive markets in the world capitalist economy. Technological innovations were initially promoted by the major industrial states as part of their military and strategic policies following the Second World War. In data processing, large U.S.-based multinational corporations have established a predominant position, with smaller electronic component and software firms competing to produce products that meet American-set industry standards. In comparison, telecommunications has been institutionalized along national lines in most of the major industrialized states, delaying the emergence of worldwide industry standards. Nevertheless, expanding consumer acceptance of electronic products combined with mercantilistic government policies patronizing national monopolies, oligopolies, and cartels have produced some staple goods in both sectors. Worldwide competition in the production of these staples has driven down the price of electronic components more quickly than had been anticipated. Rapid realization of these economies of scale has encouraged multinational corporations and the major industrial states to continue funding technoloigcal innovation and the production of customized components and software in the hope of setting the industry standards of the future.

New capital as well as consumer products and services have been produced as "spin-offs" of this research and development in electronics. Most were initially fashioned as potential substitutes for existing commodities with well-established markets. As electronics, these new products and services possessed enhanced qualities that could be marketed to some advantage. However, it was the reduction in price through competitive mass production that stimulated widespread shifts in consumer and investment buying patterns.

In the case of electronics, technological innovations were marketed in ways that did not respect traditional boundaries between industries producing different goods and services. In this case, different industries produced similar goods—for example, mechanical watches and digital watches, or mechanical as opposed to electronic toys. Satellite technology is challenging the monopoly and property rights of television networks and telephone companies within the field of telecommunications itself. New electronic product lines and production lines have evaded existing copyrights, patents, and other government-sanctioned regulations that have protected traditional capital-intensive industries from unfettered competition in the domestic market. Corporations with little vested interest in existing institutional arrangements pushed into more established industries with a zeal usually associated with small entrepreneurial firms, creating crises if not outright dislocation while contributing to the diffusion of technological change. These competitive forays sparked by technological innovations constitute the very essence of the revolutionary character of the coming information society.

It is against this backdrop of rapid diffusion of technological change in electronics industries that most prophets of the information society pause to perform an analysis of the occupational composition of advanced industrial economies. Their object is generally to emphasize the importance of the new information industries by showing just how many workers are engaged, directly or indirectly, in the production and processing of information. Most cite Porat's analysis of the U.S. economy to indicate that workers producing material commodities have been in a distinct minority since the Second World War and that most of the labor force participates in the information economy. Leiss is quite correct to refer to this categorization of the labor force as a statistical abstraction. These types of workers are not unified in any sociological sense as a political or economic force to be reckoned with. Those technical and professional occupations that are organized tend to be balkanized—more oriented toward defending claimed task domains against possible incursions by others than representing the collective interests of the greater mass of information workers.

Automation spurred by technological changes in the electronics industries is predicted to accelerate this trend toward information work, threatening in the process to disrupt the national economy. Automation in established industries will reduce employment opportunities and displace the typically male labor force in large-scale mass production and distribution. Automation in the office threatens to reduce employment opportunities among information workers as well, displacing females in secretarial and clerical jobs. Automation in foreign countries threatens to reduce the costs of our competitors, undercutting traditional export industries and even inviting imports to compete with our domestic industries. Technological transfers through direct investment by foreign-based multinational corporations will concentrate research and development employment and expertise in foreign hands and limit access of national firms to technology appropriate to national versus multinational corporate needs. The national economy is perceived to be vulnerable to technologically inspired foreign competition and domination. The ability of technological changes emanating from innovations in telecommunications and data processing to unleash competitive forays across traditional industry and national boundaries is seen as an inevitable fact of modern times. The fact that these technological and economic forces are being actively harnessed by foreign-based multinationals and their national governments is cited as a state of affairs that requires an active and concerted national (that is, government) response.

Advocates of the coming information society propose an essentially mercantilist response to the promise/threat of technological innovations in production and long distance trade. Taxation policies and governmental grants must be directed toward national corporations and firms that may be otherwise reluctant to invest in new and unconventional technologies. Large-scale investment in telecommunications and data processing must occur before worldwide industry standards are set. Such investments run the risk of becoming obsolete before costs are recovered; it is argued that the national government should cover these ventures. Where commercial and property rights are ill-defined and unenforceable (due to technologically inspired competitive forays), the national government should treat these forms of telecommunications and data processing as an infrastructure for a national information economy. Like harbors and airports and other publicly financed systems of long distance trade, these "electronic highways" would reduce the cost of doing business within the nation. Where commercial and property rights in telecommunications and data processing can be defined and enforced, they should be privatized in the hands of nationals, like railroads and banks, reducing reliance on foreign multinationals. Whether the vehicle is public or private enterprise, all these measures are designed to stimulate national investment in technological change with an eye to reducing costs and promoting the competitiveness of exports. Only gains in the balance of payments will enable the nation to deal with dislocated industries and displaced workers.

> The only strategy is one that attempts to take advantage of the benefits of technology with respect to devising new products and improving productivity. Any attempt to slow down the [information] revolution out of a concern for possible employment effects . . . would inevitably lead to an erosion of Canadian industry's competitiveness, resulting in declining exports, falling output, and collapsing employment. (Serafini and Andrieu, 1981:94)

Professor Leiss rejects this scenario, stating that this outcome is not inevitable and that the options available to the national government are not zero-sum. (I have already touched on his third point—that information workers do not constitute a dominant class in post-industrial societies.) I accept the thrust of his commentary on the arguments advanced by the advocates of an information society. To explain my reasons for doing so, I will examine the character of market competition (the revolutionary force unleashed by the information revolution) and the validity of technological determinism, a thesis that lies behind most visions of an information society.

The large-scale, technologically sophisticated sectors of society are precisely the ones least likely to be subjected to the uncontrolled whims of market competition. Unlike less capitalized sectors of society populated by relatively small entrepreneurial organizations, the more technologically sophisticated sectors are typically dominated by gigantic vertically integrated corporations. The massive volumes of goods and services that flow through these giants exceed the capacity of most world-scale markets (Chandler and Daems, 1981). This flow is coordinated by management rather than by market forces so as to maximize continuous utilization of capacities of production and distribution (Williamson, 1981). These corporations were the first to harness advances in telecommunications and data processing to avoid or minimize exposure to the short-term vagaries of competitively inspired market scarcities and gluts. Ironically technologically designed economies of scale and the massive capital outlays required to achieve these "economies" also required that the rate of technological innovation and obsolescence be subjected to management planning and control. Organizations with a large stake in established technologies and markets have an interest in slowing the rate of technological change to match the ever-increasing number of years needed to launch and amortize new products as well as new systems of production and distribution (Galbraith, 1978). It is this stake in existing technologies and systems that turns technologically sophisticated giants into technological "conservatives."

Of course, competition spurred by technological change does intrude on the world dominated by these corporate "conservatives." The spirit of "free enterprise" can still be found in the retail, wholesale, and trades-professional sectors of most industrialized and capitalistic economies. Individual firms in this competitive periphery have a miniscule stake in established technologies and products, and are not reluctant to use innovations that challenge the existing production and market arrangements of the majors. This potential is activated when discoveries threaten to create a glut in the world supply of a well-defined product or to replace a well-defined product (capital or consumer good) with a new product or production standard. Like a fox set loose among the geese, such discoveries can force the giants into a competitive frenzy. If a new product standard is being established, this initial competitive adjustment is followed by an industry rationalization, including a relocation of world-scale production and distribution facilities.

Normally the first form of competition—peripheral entrepreneurialism—prevails: the latter two forms—rapid exploitation of new discoveries and the development of radically new product standards—may be prompted by technological innovations. Large-scale, technologically sophisticated corporations are not menaced by the first and have devised strategies that limit the incidence and duration of the latter two forms of competition. The industrial giants and organized groups in the labor force will seek to commandeer and institutionalize the information revolution before it breaks out, and will be party to its containment when and where it does.

Peripheral entrepreneurialism: In most industrialized and capitalistic economies, the largest industrial corporations coexist with the smallest entrepreneurial firms. During normal times, the giants do not compete with entrepreneurs. Rather, a symbiotic relationship seems to bind these two different forms of organization (House, 1980). The majors subcontract out routine production, distribution, service, and retail work to entrepreneurs—benefiting from lower overhead and competitive bidding. The entrepreneurs rely on the majors for sustained market demand as well as for the source of experienced and technically qualified personnel. Because of the relative ease of incorporating new businesses, technological innovations that extend the productive capacity of limited resources (for example, scarce skilled labor) and that promise (only promise) potential reductions in unit costs may trigger competitively driven diffusion of new, small-scale technologies among entrepreneurs. These innovations probably will produce dislocations in the industry—bankruptcies among conventional businesses, layoffs among their blue and white collar employees. However, this revolution does not attack the large-scale systems that constitute the core of the giants. Indeed, conventional large-scale technologies may become more entrenched as the major corporations will tend to profit from any increase in technological efficiency among their subcontractors.

Rapid exploitation: Rapid exploitation of new sources of resources or rapidly opening access to new sales markets forces the existing corporate giants into contests for ultimate control. House (1980) describes the competition among the majors that follows the discovery of very large oilfields. In the course of the exploration needed to delineate the field, the majors compete with entrepreneurs and with one another to establish property rights. This rivalry subsides when (and if) the majors succeed in securing effective rights of eminent domain, regaining the ability to allocate and schedule the volume and price of crude oil production.

A similar pattern surrounds changes to the rules of doing business, changes that hold the promise of opening a major new market. Technological innovations that greatly reduce the cost of transportation have historically altered the flow of long distance trade subjecting producers to extra-local competition as well as creating a potential mass market for exporters (Marchak, 1979). However, moves by a major multinational corporation to improve access to major markets tends to provoke its peers to secure matching rights of distribution and production (Geriffi; 1978: Knickerbocker, 1973). As with discoveries, such competition abates when available production and marketing rights are controlled by the major corporations.

It is not surprising that technological revolutions set off a somewhat similar pattern of events. Paul Hirsch's analysis (1969) of the U.S. music industry in the 1940s through 1960s showed that three technological innovations—television, the transistorized radio, and the long-playing record—introduced a period of competition lasting about a decade. In all three situations—major discoveries, reduced cost of transportation, and technological innovations—the largest and most technologically sophisticated organizations reacted in similar ways to curtail unlimited competition and to limit the duration of competitively induced dislocations of large-scale systems of production and distribution.

Radically new products: In fact, the introduction of technologically novel products or production systems is associated with two competitive phases. The first involves the emergence of new relatively small-scale technologies that allow entrepreneurs to compete with the conventional large-scale technologies of the majors. Hirsch (1969) found that the radically lower production costs of the long-playing record permitted local promoters to produce records that theretofore required the facilities of a major recording studio. The cheap transistor radio so expanded the potential audience that local stations were able to gain the numbers of listeners equal to those that could only be assembled previously by a network of stations. In the ensuing competition between local promoters and major studios, and between local stations and the networks, the industry leaders were debureaucratized, forced to resort to subcontracting work previously done in-house.

The second phase occurs during the reconsolidation of the industry. Product by product, wider market acceptance comes to define new product standards (and, indirectly, new production standards as well). Given established standards associated with known markets, research and development strives to identify the scale of production and distribution systems that best achieve economies of scale. While the installation of large-scale systems generally eliminates entrepreneurs as competitors, marking the end of the first phase of competition, there is a period in which the majors continue to vie with each other.

Facilities are often relocated in attempts to lower the calculated costs of labor, transportation, and taxes (Vernon, 1966). Unlike entrepreneurs whose only control over operations profitability is through improved technical efficiency, corporate giants alter profitability through negotiations with one another, with governments, and with other sizable organized groups in the economy (for example, unions, professional associations, and industry associations). In fact, taxes and grants, rights to import and export, utility costs and government-financed infrastructure, property rights, and other ground rules regarding financing, labor relations, and other legally enforceable responsibilities are thought to have a greater bearing on profitability than technical efficiency (Commons, 1961; McNeil, 1978; Galbraith, 1978).

By regulating entry into key markets and by sanctioning copyrights, patents, and other property rights, governments permit large corporations to choke off competition. Government deregulation, or inaction in the face of radical technological innovations may threaten the viability and attractiveness of large-scale projects. While no single government may be able to assure profitability on world markets, the profitability of large-scale operations is more a matter of negotiation than of achieving greater technical efficiency. Technical innovations may have initiated the contest, but successful negotiations rather than efficiency determine its ultimate outcome. Claims of the inevitability of the information revolution are based on entrepreneurial notions of competition, notions that do not acknowledge the primacy of negotiations for the incorporation of large-scale technologies. And there is nothing inevitable about negotiations.

Calculations of optimum scales of operation and redesign of the flow of goods and services within multinational corporations and industry will prompt adjustment in the location and form of the industry. But once order is restored, the consolidated arrangements are remarkably impervious to further change. In his analysis of the composition of various U.S. industries, Arthur L. Stinchcombe has found that their structure tends to reflect the institutional context in force at the time of their last major technical reincorporation. It was not uncommon for the character of an industry to remain essentially the same for up to a century despite the flood of technological innovations that has marked the twentieth century (Stinchcombe, 1965).

Technological determinism is often presented as a Convergence Thesis. John Kenneth Galbraith, for example, suggests that *The New In-*

dustrial State with its cooperative (that is, negotiated) rather than competitive form of industry and corporate organization is seen as the product of the increasing scale of technological systems (Galbraith, 1978). There is, he claims, a convergence in the social composition and organization of society in the Soviet Union and the United States because both are adopting analogous technological systems.

Joan Woodward (1965) advanced a somewhat similar argument to explain differences as well as similarities in the composition and structure of British manufacturing organizations. Distinguishing between craft, mass production, and automated production technologies, Woodward found that each technology tended to be denoted by a distinctive style of administration. The more the composition and structure of management converged toward the technological norm, the more profitable the manufacturer was found to be. For each technology, there seemed to be "one best way." Woodward's explanation (but not Galbraith's) implied that these correlations were a product of natural selection and survival of the fittest in a competitive economy. Market competition was seen as the force behind a technological determinism, motivating managers to optimize technologically defined priorities. In the words of advocates of the information revolution quoted earlier, "the only strategy is one that attempts to take advantage of the benefits of technology with respect to devising new products and improving productivity."

Technological determinism seems to be plausible, so where is it deficient? It has already been noted that competition is not an ubiquitous and unending feature of our capitalistic economy. Competition is a feature of the peripheral entrepreneurial economy, while negotiation and other cooperative strategies are more characteristic of giant corporations. Significantly Woodward's observations were based on a survey of relatively small firms (Blau et al., 1976). Surveys of larger organizations have failed to find any correlation between technology and managerial organization (Hickson et al., 1969; Child, 1973; Blau et al., 1976). Differences in production technology do tend to affect the grouping and hence first-line supervision of production workers, but production technology does not appear to determine the consciousness or culture of the workers (Gallie, 1978), or to determine the shape and style of the administration away from the plant floor. Differences in the use of computers in industry have been found to have similarly limited effects, being associated with only a slight centralization of administration at the point or level of applications but with negligible ramifications for the larger organization (Blau et al., 1976).

Claims of technological determinism and its inevitability seem to be most apropos for small-scale entrepreneurs. In contrast, ironically, the reasoning of advocates of the information revolution appears to be least relevant to large-scale corporate technologies in telecommunications and data processing. If there is a convergence in the social organization of the economy dominated by corporate giants, this convergence probably has more to do with cooperative than competitive processes.

14

Microelectronics, The Concept of Work, and Employment

STEPHEN G. PEITCHINIS

This exploratory discourse on the concept of work and the implications of the advent of microelectronic technology was prompted by the increasing concerns about the possible onset of what is frequently referred to as "jobless growth" (Freeman, 1978)—and the possibility that microelectronic technology will facilitate the production of increasing quantities of goods and services with decreasing quantities of labor inputs.

The production of increasing quantities of output with decreasing quantities of labor input is, of course, nothing new; it has been demonstrated in all goods-producing processes, and particularly so in agriculture where output has risen four- and fivefold over the past three decades, with labor output more than halved in absolute terms. But according to prevailing economic thought, on the aggregate heretofore we have had, not jobless growth, but job-creating growth. And this, too, has been well demonstrated: in 1950, total employment in Canada averaged 4,976,000; in 1961, when the first computers were finding their way into commercial establishment, employment stood at 6,055,000; in 1971 it was 8,079,000; and in 1981 it increased to 10,933,000. Evidently the increasing use of computers and other electronic instruments over the past two decades has not affected employment negatively on the aggregate. Why then the current concern?

Work prepared when the author was a Killam Resident Research Fellow at the University of Calgary in 1982.

Increasing concern about the possibility of jobless growth is founded on the expected acceleration of microelectronization of services-producing processes. These processes have been the major source of job creation over the past three decades. Almost 90 percent of the increase in the labor force over the period found employment in them. The possibility now that these largely labor-intensive processes might become capital-intensive brings into view the experience of employment in agriculture and other goods-producing processes—very substantial increases in output with substantially diminishing labor inputs. Even if the microelectronization were to take place gradually, and not affect adversely those presently employed, the failure to take into account the annual increase in the labor force will result in substantial unemployment.

The validity of this concern depends on the interpretation of the term "job-creating-growth." What is the relation between growth and jobs? Was the heretofore recorded job creation a condition for growth or was it facilitated by growth? In other words, was all of the employment that was created over the past three decades necessary for the growth process itself, or was some of it the result of the growth, in the sense that the growth facilitated the creation of paid employment as an end in itself? As a social and political objective?

Given the possibility that growth both creates jobs that are functionally related to the production processes and facilitates the creation of jobs not so related, then the issue of concern should be growth itself and not whether the growth is accompanied by the creation of jobs functionally related to it.[1] In such a case, to the extent that microelectronic technology will increase productivity and growth, it will both create and facilitate the creation of jobs.

Acceptance of the proposition that some jobs may not be functionally related to production processes, directly or indirectly, is made difficult by the existing definition of employment and the relationship between an employed person and the national output. A "gainfully employed" person is generally deemed to contribute to the national output regardless of the nature of work and regardless of the source of payment. The question is, does it matter whether the gainfully employed person is a domestic performing work activities that would have been performed by members of the family, and sharing the family earnings; or a civil servant monitoring developments in Poland and El Salvador; or an advertising executive experimenting with a new toothpaste advertisement that will increase sales by 10 percent at the expense of another toothpaste; or a public relations person whose work functions are to develop and maintain a favorable public image for the enterprise, regardless of the nature of activities in which the enterprise is engaged? On the other hand, there are people who are not gainfully employed, and yet their activities may involve contributions to the national output. For example, activities related to home and family, hobby-work activities, activities related to social services, and such other.

The concept of work has changed significantly over time; attitudes towards work, the nature of work activities, and work-output relationships have all changed. Perhaps a new approach toward work and working is warranted, and the advent of microelectronic technology may compel the evolution of such a new approach.

THE CONCEPT OF WORK

What is "work"? Scholarly writings abound on the work ethic, work motivation, work and leisure, but not on the concept of work. It is generally taken for granted that the concept is well defined, but it is not. (See Mead, 1967; Bell, 1973; Smith et al., 1969; Borow, 1964; McClelland, 1961; and Nosow and Form, eds., 1962.) The question is fundamental at a time when it is increasingly asserted that the integration of computers and telecommunications technology in production processes will make work increasingly more scarce. But what work? The work to which reference is made in this context is usually limited to activities associated with the production of marketable goods and services. There is a problem with this. Many activities produce goods and services that bypass the market, such as those related to home and family, voluntary services rendered to churches, clubs, public service organizations, and health and welfare institutions; and equally many activities are classified as work under some conditions and not under others. For example, the activities of professional athletes and artists are deemed to be work, while the activities of amateur atheletes and artists are not so considered, even though the time allocated to the activities and the intensity and effort involved in them may be the same. What is the difference? Essentially it is that one group performs its activities for money while the other does not. The moment the amateur begins to perform for money the activities become work. This suggests that it is not the nature of activity that determines whether it is work, but rather whether a direct payment is made for the performance of the activity.

Joseph A. Schumpeter referred to technological change as the process of "Destructive Creation" (Schumpeter, 1947). It destroys products, processes, and services and creates new products, processes, and services. The organizations and structures of industrial, commercial, and institutional enterprises change, the range and nature of work functions change, and the nature and structure of occupations change. One need but reflect on the changes in goods and goods-producing processes over the past half century, on changes in transportation, distribution, and retailing, and on changes in the broad spectrum of services, to appreciate the magnitude of the Destructive Creation. Scores of occupations and employments have been destroyed, scores more have been created; and the work functions of most occupations and employments that have continued to exist have changed so much over time as to make the occupations and employments the same in name only.

The Industrial Revolution is commonly associated with the introduction of machine production. The role of labor in production processes became increasingly complementary to the machine. The electronic revolution will be associated with the introduction of the computer. Production processes will increasingly become an interface between computers and machines. The role of labor in the context of existing processes, existing work functions, and existing work relationships is uncertain. Most certainly the functional work relationships of the vast majority of the working population will change drastically. Probably more people will be involved at the beginning and the end of production processes than in-between; the work itself will probably involve more inputting the system and supplementing the outputs of the system than participating in the system itself; the worker-work relationship will likely become less rigid than it is at present; supervision and assessment of work will shift to the computer; and work-time over a lifetime will decrease significantly as will hours of work per day, week, month, and year.

Each of us has about sixteen hours of waking time per day, 112 hours per week. How is it used up? The average wage and salary earner allocates about forty hours to paid activities, and the remainder to unpaid activities and relaxation. Technology affects the allocation of time to all activities, paid and unpaid, as well as among different forms of relaxation. Note, for example, the changes recorded between 1934

Unpaid Activities	1934 Minutes per day	1966 Minutes per day
Housework	95	140
Purposeful Travel	129	76
Time Spent Shopping	16	34
Sub-total	240	250
Relaxation:		
Meal Time	107	70
Visiting	26	46
Listening to Radio	26	4
At the Movies	22	3
Reading Books	22	9
Walking for Pleasure	22	1
Playing Cards	9	4
Attending Sports Events	7	2
Correspondence	3	6
Watching Television	0	90
Sub-total	244	235
Grand Total	484	485

and 1966 in the allocation of time among certain unpaid activities and forms of relaxation (Scitovsky, 1977:163):[2]

It would appear that technology has very significantly influenced the allocation of discretionary time among activities and forms of relaxation; improvements in transportation have almost halved the time allocated to purposeful travel, despite the substantial increase in urbanization between 1934 and 1966, and the resultant increase in distances from work. Changes in the physical structure of retailing establishments, and the considerable increase in varieties of goods available on the market, have almost doubled the time allocated to shopping; television has caused havoc with the allocation of time to various forms of relaxation—meals, movies, radio, books, sports events. All these lost time to television.

The time allocated to paid work activities now stands at only about 35 percent of the total waking time available each week. Yet, people seek further reductions in work-time. This does not necessarily mean that people generally find the time allocated to paid work to be too much; it means rather that they want more discretionary time to themselves. It is quite conceivable that some or many of them will allocate some or all the additional discretionary time to paid work of one kind or another. Flexibility in the allocation of time among activities and forms of relaxation appears to be the goal, not reduction in the time allocated to work activities. If this, indeed, is the goal, then microelectronic technology has very important implications. Indications are that patterns of work will change significantly, and human participation in work processes will become considerably more flexible than within mechanical processes (Toffler, 1981; Nora and Minc, 1980; Barron and Curnow, 1979). It is widely expected, for example, that the introduction of a "telematique" infrastructure[3] will expand the range of activities performed from home considerably beyond what the telephone has facilitated. When the computer becomes the source of and intermediary for all information required for the performance of most work functions, and access to it becomes possible from any place, at any time, regular and routine visits to the place of the employer may no longer be necessary. This will save some of the time allocated to "purposeful travel," which can be used in other work activities or relaxation.

The question of "what is work" has been answered implicitly: work is any activity that is not specifically classified as relaxation. One hundred and twelve hours of waking time available to each of us each week is allocated among four groups of activity: paid work activities, unpaid work activities (housework, church work, club work, volunteer work), leisure work activities (painting, gardening, and such other), and relaxation. Technology has affected the allocation of time among the four groups and within each group in the past, and indications are that it will do so in the future. Indeed, Jonathan Gershuny (1978) suggests the possibility of some coalescence between certain household work activities and relaxation activities.

THE CONCEPT OF LEISURE

What is leisure? Aristotle (*Nichomachean Ethics*, Book X, ch. 7) distinguished between leisure and amusement. Leisure was the source of true happiness: "And happiness is thought to depend on leisure; for we are busy that we may have leisure." Amusement was for refreshment so that exertion in work would be enhanced: "Happiness, therefore, does not lie in amusement; it would indeed be strange if the end were amusement; . . . Relaxation, then, is not an end; for it is taken for the sake of activity" (Book X, ch. 6). In other words, leisure served the need to unwind, to relieve tension.

We divide leisure time into activity and relaxation. The quest for more leisure time is for more discretionary activity time, not for more relaxation ("consumption") time. The problem arises from the fact that traditionally time not allocated to productive activities, as defined by economists, has been deemed to be allocated to consumption. As John D. Owen puts it: "The definition of leisure time as consumption time emphasizes the subjective intent with which an activity is approached, rather than the objective purpose which it serves" (Owen, 1969:5). In Japan, the calisthenics in which workers engage daily are deemed part of the work activity; and in the Soviet Union, political indoctrination would be classified as participation in work activity. Most of us do our calisthenics and political learning during our leisure time. Are the activities production or consumption?

This question, of whether leisure time that enhances economic welfare should be given a value and included in national income, has been subject to considerable discussion. Simon Kuznets puts the question this way:

> Considering that human life is short and that time spent on unpleasant tasks is a negative element, should we not insist that economic welfare and progress be measured not only in terms of supply of wanted goods but also in terms of freedom from unwanted labor? (Kuznets, 1953: 208; also see Nordhaus and Tobin, 1972)

To the extent that work is a sacrifice, or a disutility, an increase in leisure time will increase economic welfare. The additional pleasure gained from the additional leisure will more than offset the increment in output lost by the decrease in paid work-time; furthermore, to the extent that the leisure time is used up in amusement (in the sense used by Aristotle), productivity during work-time will increase and offset some of the potential loss in output. Therefore, to the extent that leisure time contributes to an increase in economic welfare and output it should be given the value of work.

There is then the additional problem of distinguishing between work and leisure when the same activity is involved, and when the activity is performed with the same degree of intensity. Some people are deemed to work while training and playing hockey, football, golf, tennis, chess, cards, and snooker, whereas others, similarly engaged, are deemed to engage in leisure activity; some people are deemed to work while engaged in activities related to painting, the playing of musical instruments, and performing on the stage, whereas others involved in the same activities are deemed not to work. What is the discriminating element? It is the assignment of a value to the activity and the willingness of someone to pay that value. In a market economy, market forces assign values, and buyers determine who works and the amount of work. In non-market economies, governments, government agencies, and social institutions assign values, make the requisite payments, and thereby determine who works. Therefore, the transformation of leisure activity into work activity appears to be a matter of assignment of value and payment for the activity. Many a casual, "leisure time" painter has been transformed into a working painter by a government decision to fund a collection of paintings; the leisure activities of many musicians have been turned into work activities by government grants to orchestras; and philanthropic organizations have made it possible for many "leisure" musical, writing, and acting activities to become work activities.

All of this reveals a muddled reality about the concept of work. It is not the activity that is being defined, but rather the payment for the activity. Whether an activity is work or non-work, and whether an activity is work activity or leisure activity depends on whether someone is willing to pay for it. This means that employment, which by definition is limited to paid activities, is determined by the ability and willingness to pay for whatever activities people can perform. The critical question then is the relationship between what people do and the ability to pay. If ability to pay is independent of the activities in which most of the people engage, then full employment can be maintained by the simple act of payment for whatever activities people wish to engage in. Given social and political purpose to community existence, the activities could be stratified and payments made accordingly. This is not utopia, Communist or any other. It is a reality that is being suggested as a possible outcome of the application of microelectronic technology to all production processes. The technology is expected to produce the surpluses that will facilitate the ability to pay for activities related to and arising from the technology, as well as for activities related to general social and economic well-being. Perhaps we are on the way to concepts of work and leisure that depart significantly from those in existence at present. Margaret Mead (1967: 15) commented on the issue in the following way:

> . . . one of our central problems at present is to begin to get rid of this dichotomy between work and leisure, between what you are paid to do, the way in which you get hold of a bit of the currency of the country, and your involvement in the country—between the right to experience the benefits, at a certain level of food, medical care, and education, and the possibility of using one's gifts to the limits.

ATTITUDES TOWARD WORK AND NATURE OF WORK

Attitudes toward work have varied widely through history. What work one did and did not do, and what work was appropriate and not appropriate, have been dictated by social norms founded on tradition, economic conditions, social structures, and religious doctrines. "To the Greek mind" writer Alexander Gray reports (1951: 14), "nearly all activities of modern society would have appeared unworthy and debasing. When slaves are at hand to do the work, work becomes the mark of the slave." Similar attitudes have prevailed in socioeconomic systems governed by castes, and where large proportions of the population have existed at poverty levels. All physical work was considered appropriate for the lower class, the slave, the domestic servant, and the hired hand.

To Plato, Aristotle, and Xenophon, husbandmen, builders, weavers, shoemakers, "mechanics," and artificers of all kinds were indispensable to the existence of the state, but they could not be citizens. Citizenship was an art, requiring much study and many kinds of knowledge, which dictated that it be the primary occupation. To Cicero (*De Officus*: 153–55), most occupations were "vulgar":

> Now in regard to trades and other means of livelihood, which ones are to be considered becoming to a gentleman and which ones are vulgar, we have been taught, in general, as follows: First, those means of livelihood are rejected as undesirable which incur people's ill-will, as those of tax-gatherers and usurers. Unbecoming to a gentleman, too, and vulgar, are the means of livelihood of all hired workmen whom we pay for mere manual labor, not for artistic skill; for in

their case the very wages they receive is a pledge of their slavery. Vulgar we must consider those also who buy from wholesale merchants to retail immediately; for they would get no profits without a great deal of downright lying; and verily there is no action that is meaner than misrepresentation. And all mechanics are engaged in vulgar trades; for no workshop can have anything liberal about it. Least respectable of all are those trades which cater to sensual pleasures:
"Fishmongers, butchers, cooks, and poulterers, And fishermen," as Terence says. Add to these, if you please, the perfumers, dancers, and the whole *corps de ballet*.

But the professions in which either a higher degree of intelligence is required or from which no small benefit to society is derived—medicine and architecture, for example, and teaching—these are proper for those whose social position they become. Trade, if it is on a small scale, is to be considered vulgar; but if wholesale and on a large scale, importing large quantities from all parts of the world and distributed to many without misrepresentation, it is not to be greatly diparaged. Nay, it even seems to deserve the highest respect, if those who are engaged in it, satiated, or rather, I should say, satisfied with the fortunes they have made, make their way from the port to a country estate, as they have often made it from the sea into port. But of all the occupations by which gain is secured, none is better than agriculture, none more profitable, none more delightful, none more becoming to a freeman.

This is, of course, a commentary on the social status of occupations and employments, not a commentary on the productivities and contributions of different occupations and employments to the national output. Nevertheless, the impression is conveyed that like the "natural" and legal slaves in Aristotle's Greece (*Politics*, Book I, ch. 6),[4] the "vulgar" occupations were regarded as nothing more than instruments of production. This raises two questions: one is whether there is any substantive difference between the attitude recorded by Cicero and the economist's concept of "labor," and the other is whether it is possible that the national output can be produced without slaves and vulgar occupations, so that most of the people can be citizens and freemen.

Christianity postulated a doctrine totally opposed to the Greek and Roman views. God Himself worked, rested from His work on the seventh day; Christ worked; and St. Paul worked day and night so that he might be "chargeable to anyone" (Gray, 1951: 45). Indeed, he laid down the rule, often attributed to Marx, that any one who would not work, neither should that person eat. To work with one's own hands was dignified and noble; to labor was a duty.

Marx accepted fully the Christian dogma on the matter: to him, work was "the act of man's self creations, . . . not only a means to an end—the product—but an end in itself, the meaningful expression of human energy. . . . Only in being productively active can man make sense of his life" (Fromm, 1961). He believed that work should be the most important source of human enjoyment, but to make it so, one had to remove from it "the compulsion of economic necessity" (Scitovsky, 1977: 90). Work remains a duty, but to oneself—for self-fulfillment—and to society. Presumably, the removal of economic necessity from work would free people from all undesirable forms of work, widen the range of choices, and benefit both themselves and society at large—Adam Smith's invisible hand in the pursuit of self-interest, but without economic motivation. Remove economic necessity from the motivation to work, and you remove Jeremy Bentham's pain. Involvement in work becomes a personal choice, which means people can chose work that is more suitable to their desires, more suitable to their natural capacities; work itself becomes all pleasure. This is one of the issues to be addressed in relation to microelectronic technology. If computers, computer-related instruments, and telecommunication networks can produce the national output required for a generally rising standard of living, with the participation of a relatively small proportion of the population, the Marxian prophecy may yet be realized, even though the political setting may be somewhat different from that which he envisaged.

THE ROLE OF MACHINE PRODUCTION AND SPECIALIZATION IN WORKER ALIENATION

Today, negative attitudes toward work are commonly associated with the routine, monotony, discipline, time-scheduling, and work-process regimentation, not with any physical or mental strain (Scitovsky, 1977: 89–105, 206–8, 279–89).[5] It is the specialization in work, the dictates of machines and processes, and the work routines formulated by others and enforced by others that are the major sources of alienation. Specialization removed the worker from the product and replaced it with only a part of it, often an infinitesimal part. To paraphrase Neil W. Chamberlain, a job is no longer a specific undertaking, to be started and completed, with another to follow; rather, it is a work function that one does day after day, with little or no variation, as part of a process, but often meaningless by itself (Chamberlain, 1977: 84). Such a transformation in work activity was dictated by the quest for increased productivity, specialization being the instrument of productivity. The combination of specialization and machine production removed labor from its dominant role in the production process, took away from labor control of the production process, and most importantly, separated labor from its product (Scitovsky, 1977: 207).[6] Although productivity increased, it probably increased by less than the potential of the machine processes complemented by an enthusiastic labor force.

Management's response to this development has been critical to labor's attitude toward work and the evolution of production processes. There is no evidence in the literature of any concerted or widespread effort to reestablish worker association and identity with the product, nor is there any evidence of effort to organize production processes in ways that would establish better relationships among the workers and between the workers and identifiable segments of the production processes. Some applied the much discussed and now discredited Hawthorne method (see Rice, 1982: 70–74), and in recent years some experimental successes have been reported from Sweden and Germany.[7] But most processes remain in the traditional large-scale electromechanical infrastructure, with labor performing the complementary functional role dictated by each process. An alternative approach to which management has resorted from time to time is to cause labor to identify with the company. To the extent that the product carried the name of the company—Ford, General Motors, General Electric—identification with the company would identify with the product. But considering the nature of many standardized products, not many workers would have wanted to identify with them. Although some Japanese companies and a few German companies have been successful with such an approach, there is no evidence of widespread worker allegiance to their companies. The ultimate alternative which has gained wide acceptance is to neutralize labor's effect on productivity by making machine operations independent of direct labor participation or, where labor participation continues to be required, to make its functional activities passive to the requirements of the production processes. Fully automated systems, such as computer-aided manufacturing and integrated computer-telecommunications systems, are viewed as the ultimate in technological processes which will overcome the remaining human barrier to efficiency.

Work specialization and the placing of mechanical instruments at the disposal of labor were expected to make work not only more efficient, but also easier. That they did: work became more efficient and considerably less taxing physically; but for many workers it became mentally dulling. The work performed by workers became progressively more repetitious, undemanding, and monotonous. Perhaps one can tolerate work that is physically and mentally undemanding over short periods of involvement, but the average individual enters the labor market at the age of eighteen and remains in it forty-five to fifty years! It is difficult to sustain enthusiasm and interest in any work over such a long period, much less for work that does not generate physical or mental stimulus. In this context, it is arguable whether it would not be desirable from a socioeconomic standpoint to eliminate such jobs. But then another potential socioeconomic problem would be created, namely, unemployment. What are the prospects of replacing such undesirable jobs with more satisfying work activities? Recognition that there are monotonous, repetitious, physically demanding or undemanding jobs, and jobs that are mentally numbing, and agreement that such jobs should be changed, or eliminated, relates to the demand side of the process only; what about the supply of labor side? Can the people who work at these monotonous jobs work effectively at jobs that require imagination, initiative, and discipline? Is it possible that for a certain proportion of the population the alternative to routine and undemanding jobs is no jobs? What promises are contained in the socioeconomic system emerging from microelectronic technology?

All work activities are not, of course, of the nature just described, and the differences suggest perhaps the nature of work that is compatible with human nature. An examination of work activities and processes that people find desirable reveals four general characteristics: *independence* in the carrying out of the work tasks, whether the tasks are independently determined or are required by the nature of the activity, by the nature of the process of production, by rules, regulations, or whatever; *variety* in work activities performed day to day and week to week; *flexibility* in the carrying out of the work activities, particularly in starting them, ending time, and time continuity; and free personal *choice* in entering and staying in the work activity. Such characteristics will be found in work activities performed by unpaid volunteers; by retired people who continue to work without pay, for substantially reduced pay, or small honoraria; by independent businesspersons and farmers; and by all kinds of professional people. Warren Bennis is quoted to have said that many such people, particularly professionals, "are committed to the task, not to the job; to their standards, not their boss" (Toffler, 1981: 148). It is instructive to note that such people generally work longer hours than people whose tasks and work routines are presented by employers (Scitovsky, 1977: 93). But even in such work activities personal enthusiasm cannot be sustained indefinitely. Most organized work requires commitment on the part of those who undertake to perform it; and for most people the lifetime of commitment to work is too long for sustained enthusiasm, regardless of the flexibility of the work activity itself. Many continue to work for years after retirement, and some work as hard after as before retirement, but often involving work routines in new environments. Furthermore, the work then is freely chosen, economic necessity is removed from it as a motivating factor, there is a feeling of service in its performance, and there is the knowledge that it is being performed at will.

Another element in many such work activities that contributes to the satisfaction derived from their performance is the coalescence of work with other activities. There is no separation of work from leisure the way there is in most wage and salary employments. Often it is difficult to determine when work ends. For example, when does a farmer not work? Is a scientist relaxing or working when reading a scientific journal? Is an economist working or relaxing when reading the *Wall Street Journal* or the *Financial Post* at home, at the club, or when commuting to work? These are important issues in the definition of work, particularly when less and less work can be packaged into measurable units. It may be a nightmare for bureaucrats and administrators who prefer to identify and compartmentalize everything, but it is also a relief from regimentation for other people. The regimentation, rigidity, and discipline imposed on people at work are seldom dictated by the work processes; they are conceived by people, imposed by people, and enforced by people. They are a manifestation of lack of trust in the integrity and responsibility of people in the performance of their work tasks; some are motivated by administrative convenience; some are presumed to be based on efficiency considerations; and some are based on nothing more than established rules whose rationale has long been forgotten.

The historical record indicates very significant changes in both the nature of work and attitudes toward work. There are few craftsmen left who see their work from beginning to end, and fewer still who carry out their work for its own sake, as an end in itself. The proportion of the working population involved in the production of goods has fallen steadily over time to around 30 percent in advanced industrial-commercial economies, and most of those among them who perform work activities related to the production of industrial goods have largely lost contact with the final product. The remaining 70 percent of the working population perform functions that are grouped together under services. Some of these are related directly to the goods processes and are deemed a necessary part of those processes. Transportation services, distribution, storage, repair and maintenance, and, of course, directly related office and managerial services are of this nature. Then there are services that are rendered directly to the people at large, such as personal services, educational and health services, and financial services. All of these are identifiable, and work functions performed can be related to specific purposes. But there is a third group of services which are not related specifically to the production, distribution, and consumption of goods or to the people at large. These are found throughout the economy and society, in most large establishments—industrial, commercial, and institutional. These are the work activities that often elicit the question "but what do you really do?" They are difficult to define, difficult to measure, and difficult to relate to the national output, as if they have been created because of ability to pay or to justify payment. These are the employments that yield zero marginal product and are paid from the marginal products of other factors of production.

The issue of whether a certain proportion of total employment is not related to the national output in a functionally productive way is fundamental to the question of employment in a microelectronic production system. If some employments are not related to the national output in a productive sense, then such employments have been created not as instruments of production, but as mechanisms for income distribution.

THE ECONOMIST'S CONCEPT OF WORK

The economist's concept of work is limited to those activities that are performed solely for an actual or implicit payment. People are deemed to work, in an economic sense, if they are working for pay or profit, or their work is associated with a farm or business enterprise operated by a related member of the family. Implicit in this definition is the assumption of a contribution to output. Economic logic dictates that an activity cannot be considered work if it does not contribute to the output. When payment is made for an activity, it is implicitly assumed that the payment reflects the contribution of the recipient to the value of the output. If the recipient has made no contribution, then the payment will be at the expense of someone who has made a contribution or at the expense of one of the other factors of production.

This presents a *problem* and a *possibility*. The *problem* is with the identification and measurement of contributions to output. With goods, and the labor employed directly in their production, there is no particular difficulty; the same can be said to some extent about all marketable services. But in many instances it is very difficult to relate individual activities to the output, and measurement of the contributions is doubly difficult. There is then the problem of deciding which contributions enter the output and which do not. When payment for an activity is the criterion determining work and by implication contribution, then the act of payment itself becomes the determinant of the contribution, not the activity that is being performed. Thus, the valueless activity, in an economic sense, of an amateur acquires value upon the payment of a salary. For example, the valueless activity of a hospital volunteer acquires value upon the payment of a wage; hence, the consequent change in classification from volunteer to employee. The activity remains unchanged; yet the output associated with the activity is regarded as a contribution to total output under one classification and not under another. As long as a political functionary renders services to the party organization on a voluntary basis he or she is deemed not to contribute to the national output; but the moment the party gets hold of some funds and begins to pay for the services, the services become work, and the payment is considered to reflect a contribution to the national output.

The *possibility* is that many of the activities for which payments are made, and as a consequence are thought to be work activities, do not in fact make any contributions to output. The creation of such activities is determined by ability to pay; they do not determine ability to pay. Payments for such activities are made from the contributions to output of other work actiities and the contributions of the other factors of production. This possibility finds partial support in the fact that over time there has not been any significant change in final expenditures for services as a share of total expenditures, nor has there been any significant shift in the composition of real output from goods to services.

The respective sectoral shares of Canada's gross domestic product have remained the same since 1951 (Economic Council of Canada, 1979: 73). Such evidence casts doubt on the often repeated assertion that the rapid increase of employment in the service sector of the economy is the result of increases in demand for services as incomes rise.

Economists make reference to three factors of production: land (natural resources); capital (machinery, equipment, factories, offices, and other produced instruments of production); and labor (everyone who participates in paid work activity—from the chambermaid to the President). The role of technology is conceived as an element in production processes which increases factor productivities and facilitates an outward movement of the production frontier with given factor quantities. This is most significant to the subject of our argument: it suggests the possibility that the national output can be increased with decreasing participation by labor inputs. This is a historical reality, of course, manifested in agriculture, in manufacturing, and in the very substantial decrease in manhours per unit of output. The critical issue is the proportion of the national output required to maintain an increasing productive capacity in the economy. In other words, how much of the output can be taken away from the factors that participate in its production without impairing their productive capacities? That amount will determine the payments that can be made and activities that can be created in industrial, commercial, and institutional enterprises in response to organization, social, and political considerations. Acceptance of this proposition will give implicit recognition to what I believe to be a reality, namely, that, although all employment is related to output in a distributive sense, all of it is not so related in a functionally productive sense. Some employment is created in industry, commerce, and institutions for purposes other than its contribution to output.

The proposition is not fundamentally different from the Marxian and physiocratic views on the production of surplus value (Marx, 1969). But it differs significantly with respect to the sources of surplus value and the distribution of it. Marx attributed the surplus value to labor alone and had it taken away by the capitalist, whereas the physiocrats attributed it to the land and those involved in agricultural work activities and had it taken away by the landlord. The proposition here is that the surplus output is the result of the application of all factors of production functionally associated with the production processes. The allocation of the surplus is for the most part directed toward the attainment of social and political objectives, one of which is the creation of employment.

The association of employment with output in a functionally productive way has had a limiting influence on efforts to create employment. The work activities to be created had to be justified on the basis of their contribution to output, and payment for the performance of the activities had to be justified on the basis of the expected contributions by the person who performed the work activities. Such determinants of work, pay, and employment left many potential social service activities not undertaken and many people not employed—not because they could not perform some activities, but because the activities could either not be justified on the basis of standard economic value-added criteria, or because the minimum payments required by law, custom, or contract could not be justified on the basis of expected contributions to the value of output. Hence, questions have arisen as to whether it is justified from the standpoint of net economic welfare to limit work activities to those only that are functionally related to processes producing goods and services. Should we not, as Margaret Mead put it, "devise a system in which every individual's participation in society is such that he has dignity and purpose, and the society has a rationale for distributing the results of its high productivity" (Mead, 1967: 10)? Whether we should, indeed, whether it would be possible to design, implement, and sustain such a system, will depend on whether the "high productivity" can be attained again and sustained over time. To the extent that microelectronic technology will make that possible, it would then be possible to create a system of participation in social and economic activities based on both functionally productive and distributive criteria.

CONCLUDING COMMENT

Whether computers, computer-related instruments, and telecommunications technology are ushering in a leisure society remains an issue of uncertainty at this time. There is increasing evidence of very significant reductions in the labor intensity of production in both goods-producing and services-producing processes. Goods and services are being produced with labor inputs as low as 20 percent of the quantities used prior to the introduction of electronic processes, and electronic instruments are being produced with as little as 5 percent of the labor inputs it took to produce their electromechanical counterparts.

Another characteristic of microelectronic technology which suggests very significant implications for employment related to the production process is its pervasiveness. Unlike electromechanical technology whose application was largely limited to goods-producing processes, microelectronic technology can be applied to virtually any known activity. As a result, it is most unlikely that the historical cycle in which a change in technology is followed by rising incomes and rising aggregate demand, and then by another round of increases in employment, will be repeated. The production capacity of the technology is so vast, its potential efficiency so great, and its pervasiveness so comprehensive as to make it virtually impossible to offset through increases in aggregate demand, even over the longer run. New goods and services will be produced, as in the past; some existing goods and services will be produced in much larger quantities; some new occupations and employments will emerge, as in the past; some of the work functions of existing occupations and employments will change; and some existing occupations and employments will increase in number. But it is highly problematical that these developments will offset the expected decrease of labor intensity in most production processes in the economy at large.

This suggests a general decrease in the amount of labor required for the increasing national output of goods and services. Other things being equal, this means massive involuntary unemployment. There are, of course, other offsetting elements: the period of lifetime work will likely shorten as incomes increase (Morse and Gray, 1980), and the continuity of work over a lifetime will likely be interrupted more frequently than heretofore.[8] The pattern of work common to the professoriate will likely spread to other occupations and employments. In addition, the total hours worked over the year can be expected to decrease substantially because of increases in holiday periods, maternity leaves, educational leaves, and reductions in daily and weekly hours of work. But there are natural limitations to these. At the extreme they are life without work. Therefore, the ultimate measure is the suggested broadening of the definition of work to include all activities that contribute to the increase in net social and economic welfare. Given recognition of such activities as work activities, by the assignment of values and the payment of stipends, employment, as defined, can increase significantly. The limit to the creation of such employment will be the rate of economic growth, the share of output that can be taken from factors of production without impairing the productive capacities of production processes, and society's tolerance of income redistribution.

NOTES

1. The proposition that some jobs may not be functionally related to production processes and that their creation and continuing existence depends on growth is not new. Mercantilists (sixteenth and seventeenth centuries) and physiocrats (seventeenth and eighteenth centuries) discussed the issue in the context of "productive" and "unproductive" labor. See Taylor (1960).

2. Time-budget data of average wage-salary earner in the United States. J. P. Robinson, "Social Change as Measured by Time-Budgets," paper presented at the 1966 Annual Meeting of the American Sociological Association. Cited in Tibor Scitovsky (1977): 163.

3. The term "telematique" was coined by Simon Nora and Alain Minc to indicate the confluence of computers and telecommunications technology.

4. Aristotle makes reference to natural capacities in his discourse on slavery. He distinguishes between "natural slaves" and those who "contrary to the intention of Nature . . . possess the bodies (but) not the souls of free men . . . ". *Politics*, Book I, Chapter 6, Welldon translation, p. 13.

5. Scitovsky (1977: 89–105, 206–8, 279–89) provides a fairly comprehensive list of studies on the issue, carried out largely by psychologists and sociologists.

6. Scitovsky (1977: 207) comments that only artists, writers, professional people and a few craft occupations still get satisfaction out of their work. "The majority of mankind has been alienated from its product."

7. The Hawthorne experiments involved changes in working conditions to increase productivity. They were carried out at Western Electric's Hawthorne plant near Chicago, from 1924 to 1932. Recent investigations of the research methods used and the conclusions reached on the basis of the evidence cast serious doubt on the validity of the reported results.

8. A recent U.S. study on retirement among professional, managerial, and technical workers found that among those who retired in 1968–69 only 13 percent retired before age sixty. Among thsoe who retired in 1976–77, 30 percent retired before age sixty. See Morse and Gray (1980).

15

Counterpoint: The Sociology of Jobless Growth: Some Reactions to the New Concept of Work and Employment

DONALD L. MILLS

If one were prepared to accept the explicit as well as implicit visions and assumptions found in Professor Peitchinis's paper, much of what he argues follows imaginatively and logically. Indeed, the major thesis, central concepts, and supporting arguments are presented with real breadth and erudition. If one were willing to agree with this view of microelectronic technology and society, there would be little for the reviewer to do except concur and embellish this utopian vision. However, many people will not be willing to grant the premises, as they seem to depart unacceptably far from the social realities of the past and present, and the likely prospects for the future.

Peitchinis defines work as "any activity that is not specifically classified as relaxation." This psychological formulation implies that what may be one person's relaxation may be another's work, and thus encourages conceptual mystification and methodological pitfalls. As an integral part of the "paid work," "unpaid work," "leisure work," "relaxation" formulation, Peitchinis says the "transformation of leisure activity into work activity appears to be a matter of assignment of value and payment for the activity. . . . (As) employment . . . is determined by the ability and willingness to pay for whatever activities people can perform . . . (with the application of microelectronic technology) full employment can be maintained by the simple act of payment for whatever activities people wish to engage in." The potential employment problem can be solved merely by redefining leisure activities—although this seemingly "simple act" assuredly involves changing fundamental social values, individual rights, and responsibilities.

Professor Peitchinis contends that there are several desirable work activity and process characteristics to be anticipated: greater "variety" of activities as where the span of monitoring tasks under worker control is broader, together with "independence," "flexibility," and "choice." But is this last-mentioned threesome really to be found? Edward B. Harvey reports that in petroleum refining and chemical manufacturing enterprises which lend themselves well to automation, "specialization ensues" (1975: 113) and it is difficult to imagine much scope for variety. Given the greater process complexity and enormous capital investment of microelectronic technology—$50,000 to $500,000 per unit in Canada (Hutchinson, 1982)—how much "independence in carrying out work tasks" or "flexibility" as to the timing of work activities is actually permissible? Margaret Butteriss's research found heightened inefficiency, activity inflexibility, low moral and job satisfaction associated with the inauguration of office automation in a corporation head office in England (1981: 6). A study by Gunilla Bradley (1981: 8–9) of operating departments in a Swedish manufacturing concern discovered similar consequences. In addition, given our ever-present Canadian concern with economies of scale in production, can we expect greater latitude in job choices?

Of course, Professor Peitchinis is pointing more to the indirect consequence of microelectronic technology in its displacement of workers and their subsequent redirection to more meaningful activities now enjoyed by unpaid volunteers as these sometimes afford greater variety, flexibility, independence, and choices. However, it may be too tempting to overdramatize people's enjoyment of volunteer work in the welfare, recreation, education, and health fields. As for the further examples of independent business people and farmers, it is difficult to think of enterprises more fraught with risks and anxieties. In addition, the assertion that "after retirement . . . work then is freely chosen, economic necessity is removed from it as a motivating factor . . . (with a) feeling of service in its performance, . . . (and a) knowledge that it is being performed at will" is an overly idealized description of those so-called retired people in Canada who are in fact partly the working poor who cannot afford to live on their pension and must take whatever service jobs they can get.

It must be recognized, of course, that Professor Peitchinis is well aware of the overall employment/unemployment implications of microelectronic technology when he cautions that, despite its pervasiveness for both goods and services, "it is highly problematical that these developments will offset the decrease of labor intensity in most production processes in the economy at large. This suggests a general decrease in the amount of labor required for the production of an increasing national output of goods and services. Other things being equal [he says] this means massive involuntary unemployment." Precisely. It is this matter that demands detailed attention.

He further suggests broadening the definition of work to include all activities that contribute to increased net social and economic welfare "by the assignment of value and the payment of stipends." The old cliche about the impossibility of legislating social change is at least a half-truth here: while it might be possible to legislate the "payment of stipends," widespread attitude changes signalling the reassignment of value is quite something else. Peitchinis concludes by saying that "the limit to the creation of such employment will be the rate of economic growth, the share of output that can be taken from factors of production without impairing the productive capacities of production processes, and society's tolerance of income redistribution." This is quite a statement. What are some of the socioeconomic realities, however?

Aside from some unionized workers and monopolistic professionals, there has been but limited income redistribution in this country during recent decades. Consider the specific examples of Alberta, Manitoba, and Ontario physicians who restricted their services to redress what they perceive to be the injustices of the 1970s when the gaps between their incomes and those of other workers were reduced slightly. Despite advances in microelectronic technology, occupational and industrial groups will do anything to maintain their historical positions of social power and monetary reward. Changes would have to come from our legislators who in ostensibly representing the public interest have typically resisted actual redistribution—except for themselves.

Present-day microelectronic technology applications do not lead me to embrace the Peitchinis thesis proposing a fundamentally restructured society: the vision is too visionary. Consider the following evidence and observations: based on their study of electro-data processing, word processors, and computer personnel in small firms and local government, D. Lamberton and his associates report (among other things) "less effective labor input from the existing work force; . . . substitution of lower for higher skills; . . . reorganization of work tasks (e.g., with a preponderant displacement of female workers); . . . growth in alienation; . . . (and a) gradual perception of social costs that are then seen to require further development . . . " (Lamberton, 1980: 4). Peter Elson (1982: 14) reports that "low back pain, creeping obesity, muscular dysfunction, and increased risk of heart diseases, are some of the chronic effects," according to Dr. Mike Cox, senior exercise physiologist with the University of Toronto's fitness unit. Cox also puts "boredom, stress, eye strain, low morale and decreased job performance" on the list. As these findings and others are seemingly at variance with the Peitchinis argument, we should examine some of these topics further.

George Ritzer reports that microelectronic technology contributes to more specialization, and for this reason turnover, tardiness, and absenteeism in the workplace are less tolerable than heretofore. Moreover, office automation brings greater individual worker accountability as errors in the overall system are less permissible, and this leads to worker feelings of heightened vulnerability (Ritzer, 1977: 239). Research at the Hans Selye Canadian Institute of Stress has noted that the very possibility of introducing office automation brings about damaging threat reactions among workers, and productivity may be affected accordingly (Frank, 1982). Bradley (1981: 13–14) found a number of serious stress differentials in a comparison of insurance company computer terminal and non-terminal users in Sweden. Workers so affected cannot directly sense the process in which they are engaged, output is strictly standardized, and they are seldom free to individualize their efforts (Gendron, 1977: 143). With the large capital investment in microelectronic technology in industrial settings, employing organizations are inclined to initiate shift work so as to optimize hardware utilization (Ritzer, 1977: 238–39)—thus introducing new organizational demands to workers who have not experienced shifts previously.

Although in some processing industries and totally automated factories there is greater individual work control and a consequent decline in alienation among production personnel (Ritzer, 1977: 26), in the far commoner Canadian situation of limited microelectronic technology applications to traditional assembly-line factories (industrial robots were used by only 3 percent of Canadian Manufacturers Association 1982 Survey respondents—Hutchinson, 1982) work pace control by individuals decreases (Faunce, 1968: 46). Edward Shils (1970: 325) reports that "the inability of the foreman to get more work out of the operators by his personality, leadership and human relations skills or by his mechanical abilities which in the past kept the machines running, has resulted in negative changes in his relationships with subordinates." Batch production technology involving computer-integrated manufacturing systems studied by Melvin Blumberg and Donald Gerwin in four firms showed "structural and functional problems for quality control, accounting and maintenance departments" (1981: 4); in addition, "most workers see themselves as not having very much control over their jobs" (1981:4).

To choose a nearby example of employer reluctance to innovate in the way outlined by Peitchinis, Terrence White's recent province-wide survey of 320 organizations is informative. Although there have been extensive applications of microelectronic technology in the oil, gas, petrochemical, and associated enterprises, only 18 percent decentralized their decision-making, only 14 percent introduced job enlargement, only 11 percent adopted job-enrichment, and only 6 percent utilized semi-autonomous work teams—this after more than a quarter century of exposure to the latest in microelectronic technology in a socioeconomic climate of affluence and state-of-the-art trendiness (White, 1979b: 22).

Let us contemplate several different but related problems to better prepare ourselves for the challenges of microelectronic technology. William Faunce (1968: 66) has warned that the employment repercussions of microelectronic technology affect segments of the labor force differentially, with the very young unskilled persons entering the labor force for the first time suffering the greatest impact of all. Furthermore, speaking of "manpower problems of major dimensions," Seymour Wolfbein notes the need for "retraining of displaced workers,..effective preparation of young people for the changing economic and technological world," together with more current and comprehensive monitoring of microelectronic technology applications (1970: 77–78). How well are we doing in these training and research efforts? The longstanding inadequacy of workforce training and retraining programs in Canada is notorious throughout the industrialized world. Federally funded retraining in Canada now involves expenditures of about $2,500 per person in-program; given the recent announcement that the federal government is to allocate $140 million for this purpose, optimistically we will be able to train less than 5 percent of today's unemployed, to say nothing of tomorrow's technologically displaced. Given our lengthy history of federal-provincial squabbling over education and training rights, the prognosis is unpromising for coping better with the Brave New World of microelectronic technology.

In yet another sphere, it seems curious that Professor Peitchinis has made so little of the very differential and immediate impact which microelectronic technology is having and will have on the circumstances of women in the labor force; it is they who are doing nearly all of the clerical and routine decision-making which is so well suited to microelectronic technology usages (Darroch, 1980: 43). According to Lucie Brunet (1980: 17), Canadian data show that between four and six jobs are eliminated every time a word processor is installed in an office—a ratio echoed in the public service from 1974 to 1980 (Ferguson, 1982: A20).

In 1980, 8.4 percent of the female labor force was unemployed, as contrasted to 6.9 percent of the male labor force (both higher now), indicating the well-known greater vulnerability of women workers. What is the likelihood of microelectronic technology spearheading a fundamental redefinition of women's status? As reported by Jim Steinhart (1982: 122), even the economic slowdown has not slackened the pace of microelectronic technology purchases by businesses and institutions "[and] there are 27,000 to 30,000 word processing stations with video screens in the country." Given the above-mentioned Brunet estimates—and using the conservative figures—this suggests that thus far these installations alone have eliminated the need for more than 100,000 jobs in Canada.

With regard to job elimination, the Canadian Press reported "that a West German study predicts that 40 percent of office work could be eliminated within 10 years, a French report forecast that 30 percent of bank employees could be redundant during the same period, and British research foresees 25 percent unemployment by 1990" (*Calgary Herald*, 1981). According to a Parliamentary Task Force, "the 80's will show a further large increase in women joining the labor force . . . [but] disruptions in offices due to the increasing use of microelectronic equipment will also affect women and create further imbalance" (1981: 3). It also warns that, "with the rapid introduction of microelectronic technology and word processors, there could be increased unemployment among women who work in offices. It is therefore imperative that governments plan special programs for the retraining and upgrading of these women" (1981: 100).

In Australia, the trade unions are concerned that consumer purchase scanning in the retail trade "may further erode full-time employment . . . and even reduce the number of casual jobs" (Lansbury, 1980: 31–32). During the 1950s, 1960s, and even well into the 1970s, there were relatively few examples of disemployment as many "replaced" workers were absorbed elsewhere in their employing organization. But "how many others do such people in turn displace or prevent obtaining employment?" as Lamberton puts it (1980: 2).

We have been considering how prepared we are to deal with the potential and problems of microelectronic technology. What about status change? Ritzer (1977: 238) discusses the reduction of status among white collar clerical workers with the introduction of microelectronic technology because of job redesign and elimination, and as noted earlier, Butteriss (1981: 13–14) encountered similar office automation problems in Britain. Overton H. Taylor (1981: 4) acknowledges that "computer operators are usually low status members of E.D.P. [electro-data processing] departments. . . . In many organizations, computer operators have low morale and high turnover and absenteeism." But perhaps these matters are not as pressing as others. What can society redefine as being a fitting work for the unskilled young people mentioned previously? Our young people have an ambiguous status; consequently we have not even managed competently with our present-day "transfers" from school to work (Hall and Carlton, 1977: 257–73). At the other end of the age spectrum, "seniors" experience a status marginality as well—"out of sight, out of mind." So-called early retirement is of little interest to many people and difficult to carry out because of recent court decisions on "ageism" and government turnarounds on retirement age. Is society really willing to provide adequate pay to "seniors" who wish to make contributions through active employment? As people on old age assistance constitute a disproportionate share of the destitute and "working poor" in Canada, what is the likelihood of providing more generously in a microelectronic technology future?

And how are we to change fundamental social values about the monetary worthiness of homemaking? Or what is now called "volunteer work"? A very substantial proportion of the population has not even come to accept the appropriateness of social welfare workers, let alone those in recreation.

It is truly encouraging to see an economist paying careful attention to non-economic variables while examining the promise of microelectronic technology. Theoretically a basic redefinition of what constitutes work-remunerable activities is noteworthy. Professor Peitchinis fully appreciates that, although microelectronic technology owes much to the wonders of modern engineering (and its scientific underpinnings), the long-term resolution of certain social consequences of microelectronic technology, as he has outlined, must depend on *social* engineering—or some form of fundamental social change. Disturbing socioeconomic dilemmas will remain with us indefinitely if we think all that is needed is technological breakthroughs.

Alienation does not inhere with microelectronic technology per se, but in how microelectronic technology is put to use—the social organization in which it is embedded (Rinehart, 1975: 22–23). For the most part management ultimately make these decisions (with a little help or hindrance from their "friends" in government), not computer designers or systems engineers! It is owner-managerial decision-making that must be scrutinized and reshaped because it has traditionally been guided almost exclusively by goals of efficiency, social power, and profitability.

New hardware has not been nor can it be the ultimate answer to implementing a better redefinition of work. Employing organizations must place the human factor at the planning apex. Appropriately redirected social values and reconstructed socioeconomic arrangements (Rinehart, 1975: 22–23) constitute essential linkages in this process; otherwise "jobless growth" and other highly injurious aftereffects of microelectronic technology will not disappear.

16

Did You Ever Meet a Payroll? Contradictions in the Structure of the Appropriate Technology Movement

ALLAN SCHNAIBERG

THE SOCIAL PRODUCTION MODEL OF APPROPRIATE TECHNOLOGY

Following the unprecedented legislative and normative successes of the environmental movement of the late 1960s, a variety of changes in the movement have occurred. Among the most important of changes have been the incorporation of energy problems into the concern of environmentalists, and the incorporation of rather dramatic new theories and programs into proposals for production reform.

Central among the proposals has been the agenda introduced by the Appropriate Technology (A-T) movement of the 1970s. This paper will focus primarily on the work of E. F. Schumacher (1973, 1977, 1979) and the organizations and theories associated with him. Much of the argument here will apply as well to the thesis of Amory Lovins (1976, 1977) regarding the necessity and social desirability of a transition from "hard energy paths" to "soft energy paths." While the ideological and theoretical origins of the two related schools of thought are somewhat different, there is sufficient overlap between the two (see Schnaiberg, 1982a; Morrison and Lodwick, 1981) to justify treating them as a single "appropriate technology movement."

At the core of the ideology of the A-T movement is the statement that in order to achieve a peaceful and permanent solution of the "problem of production," advanced societies need a substantial reorganization of production. This is necessary because of energy and related natural resource limits, and also because of the growth of alienation following the centralization and capital-intensification of modern industrial production systems. The movement sees both of these as constraints on the continuity and/or expansion of these present production systems, and envisions substantial social conflict and ecological damage (and still further social conflict resulting from the problems of allocating consequent resource scarcities) as the necessary results of extending such high-energy, high-capital systems into the indefinite future.

Rather than further reliance on this form of neotechnics (Mumford, 1973), the movement advocates a shift to a system to "accessible" technology, A-T, and/or "soft energy paths (SEP)." These would be decentralized, labor-intensive, less complex; they would provide more work satisfaction because work would not be fragmented or divided. Because of the intrinsic satisfactions of A-T work, human needs would be met more readily, and the consumption of goods and services would not need to be as high as at present because it would only have to meet basic needs and not act as compensation for the degraded work environments in the modern industrial or institutional enterprise (Braverman, 1974). Social conflict would thus likely be reduced, since stratification would be attenuated and more basic needs would be met directly from work and work-related activities. Competition for resources would likely be restricted, and natural resources would be directed toward a utilitarian end: providing the greatest good for the greatest number, through extensive use of human services and/or public/collective goods.

While critiques of the A-T ideology have been numerous from proponents of the current high-technology industrial system, they have been far less forthcoming from social scientists. In general, social scientists have either ignored the A-T movement or have tended to provide relatively uncritical support for the initial ideology and movement, as one more challenge to existing inequitable systems of resource allocation (cf. Morrison and Lodwick, 1981; Frahm and Buttel, 1980). This paper attempts to extend the review of A-T ideology and programmatic agendas in a somewhat different dimension: the implied organization of work. Specifically it raises some serious questions about the nature of labor under A-T models, questions stemming from fundamental social theories and from empirical research about some early organization of A-T production in the United States.

The title of this paper initially poses a question that industrialists frequently raise, in order to dismiss A-T ideologies: *viz.*, have A-T proponents had any experience in organizing concrete work goals around real workers—have they ever met a payroll? While this is at root simply a counter-ideology to offset the persuasiveness of the A-T claims, it also raises some fundamental social scientific questions about the dimensions of work (Rodgers, 1978; Rothschild-Whitt, 1979). This paper begins by presenting a broader social scientific conceptualization of work than A-T provides, and then it turns to some implications for the structural ambiguities or contradictions in the A-T movement's goals and means for implementing "A-T." These highlight the unprecedented role of A-T as both a distributive or equity movement, as well as a movement seeking to direct (or help direct) productive or efficient organizations which transform natural resources into social commodities. By comparison with previous *equity movements* and the *management of production*, some of these strains or contradictions in the A-T ideology can be elucidated. At the minimum, these ultimately mandate some serious reformation of A-T thinking to take account of the complexity of social reality.

THE DUALITY OF WORK: COSTS AND BENEFITS

At the most fundamental sociological level, the conceptualization of "work' in A-T writings is extremely truncated. Indeed, the focus of A-T work has rarely been on the content of work or labor, but on the concept of alternative production systems. As David Dickson (1974: 152–73) has pointed out, this type of approach indicates nothing about the relations of production, but simply starts with assumptions from the forces of production (Cooper, 1973).

If we are to comprehend the relations of production, we must first insist on a broader conception of work. A-T writers talk about work as a social right: for example, they focus intensively on having an "accessible technology."[1] By this, they mean that workers *can* control technical processes, *can* easily repair the equipment, and *can* comprehend the production processes. But the obverse of this notion is: that under A-T production, workers *must* constantly make control adjustments on technical process, *must* frequently repair equipment, and *must* learn all facets of their productive organization in order to function as productive workers. They *must* be attentive, energetic, and diligent in order for A-T production to produce.[2]

In short, what is left out of A-T ideology is that work entails obligations as well as rights, as is true of any social role. As elemental as this may seem, this duality is not incorporated into any A-T writings (cf. Ozark Institute, 1978). Thus, we fail to find any topics that arise from the history of productive industrial (or pre-industrial) societies; for example, social control of work, worker motivations, reward structures, or recruitment and training (Rodgers, 1978). While this is a political-ideological point that industrial critics make—*viz.*, that workers will perform like an ant-hill—A-T theorizing contains a serious analytic omission of the implications of work-as-obligation (quite apart from the normative commitment to A-T goals).

From a social scientist's perspective (a non-Marxist one particularly; cf. Braverman, 1974), part of the history of high-technology substitution of capital equipment for human labor is based on the distaste of workers for hard manual and mental labor (Rodgers, 1978). Automation was seen not only as profit-enhancing, but often as smoothing relations with organized labor because of reduced physical-emotional demands on individual workers. Neo-Marxist analysis (for example, Braverman, 1974; Noble, 1979) claims that high technology was introduced to enhance profitability and social control and predictability of the *labor processes* in production. Even if we accept this premise, it is not inconsistent with the previous statement. Workers chafed at early industrial labor because it demanded a great physical (and often emotional) obligation to employers on the individual workers' part. Organization of work was often designed to reduce these obligations—by changing either the content of production obligations or the time workers were obligated to perform them for a living wage (Rodgers, 1978). Moreover, automation was resisted not necessarily because it "degraded" work but because it reduced the opportunities to earn a living wage, *viz.*, gainful employment. While some resistance to work degradation had been noted, far more of the history of organized labor has been directed to maintaining employment opportunities, enhanced benefits, and reduced work-weeks (Schnaiberg, 1980: ch. 5).

Thus, the history of American productive organizations can be read in part as reflecting the resistance of workers to work obligations.[3] In contrast, the assumption of A-T is that workers seek such obligations. A-T proponents perhaps derive this idea from the notion that educated intellectuals find factory work degrading and that there is substantial demand for increased employment opportunity on the part of the unemployed and underemployed in the United States. These observations suggest a more refined statement about the social desirability of A-T-like work: *some* workers desire *some* challenge in their work, and *most* workers prefer employment to unemployment or underemployment, because the economic and social benefits of employment outrank those of unemployment/underemployment. If these arguments are correct, two basic questions arise: (1) how have participants in the A-T movement been able to ignore this duality of work? and (2) what are the macrostructural consequences of this duality for the future of A-T?

THE ROOTS OF A–T BELIEFS ABOUT FUTURE WORK ORGANIZATION

In addition to the intellectual beliefs about creative work that may underly current A-T thinking about the benefits of work, I would like to argue that the rise and form of A-T movement organizations also provide some grounding for ideologies about work. In this section I examine the nature of contemporary A-T "production" (for example, Rothschild-Whitt, 1979) in the United States, and the problems of projecting from this situation to a future A-T society.

As with most other social movements, A-T movements created organizations to articulate movement demands. In common with other movements, many of these demands center on claims for social equality, and A-T organizations have begun to frame these demands in ever more concrete terms. Many of these demands have been articulated in the form of proposals for redistribution of the rights of access to production activity, together with the production of documents and pilot projects incorporating technological innovations to provide such access. These include various forms of agricultural technology, energy production, and small-scale manufacturing within developed countries such as the United States (Pitts, 1981). What distinguishes A-T movement organizations from most other non-retreatist social movements is indeed this creation of actual production facilities, in addition to proposals for distributive justice.[4] That is, A-T movement organizations are relatively unique in providing both social pressures for welfare goals (redistribution) and specific production means (physical technology) to help accomplish these goals. Virtually every other social movement that has existed in the United States has limited its activity to mobilization of support for redistribution, but has focused such support on organizations/

agencies outside itself to implement this redistribution within the production system. This is the case with civil rights, anti-poverty, and women's movements, for example, as well as the environmental movement (cf. Rothschild-Whitt, 1979).

In the language of neoclassical economists, then, A-T movement organizations have a joint production function, incorporating both equity and efficiency goals and means. As noted later, this not only distinguishes them from social movement predecessors; it also differentiates them sharply from other production (efficiency) organizations. Presumably the initial "success" of A-T movements in creating such organizations has increased its confidence in the potential of American society to reproduce such dual-function organizations until this model permeates all of social production. But two factors appear to call into serious question this type of projection: (1) the history of A-T "production," and (2) the projected transformation of A-T from a dissenting social movement to a dominant social principle in the projected future society. Both of these limit the projections about the social benefits of productive work under A-T and thus call into question the simple models of diffusion of A-T currently argued.

Turning to the first point, we must understand that the motive for contemporary work in A-T organizations is a mixture of idealism and self-interest, of avocationalism and vocationalism (Rothschild-Whitt, 1979). A-T "workers" include movement survivors from a variety of past social movements, including not only environmental but also anti-racist, anti-poverty, anti-sexist, and anti-war ones. Many of these are participants from the inner "orbits" (Hornback, 1974) of these past (and ongoing) movements, whose motivation consists of the desire for progressive social change, for social equality and distributive justice. Some of these hold paid employment outside A-T, so that A-T "work" is typical movement work, not vocationalism. Others work for small wages, a type of "idealistic vocationalism" (which may be a brief portion of their adult lives, in fact, if history is any precedent).[5] In addition, a small portion of A-T workers are real workers: those motivated by the need for wage income primarily, for whom the work *in* A-T organizations is a job. They may or may not be attracted to the goals of A-T, but that is more peripheral to their work motivations. Interestingly we rarely get any information about or from such participants in A-T production, but from the two previous categories of "workers": these constitute our images of A-T production (cf. Rothschild-Whitt, 1979; Pitts, 1981).[6]

To address the second point, how can we *project* from this current structure of "deviant" A-T production to the future of "dominant" societal A-T production? Does the fact that small-scale A-T enterprises have "worked" imply that a society-wide transformation is likely and/or feasible? Following Gunnar Myrdal (1968), it would appear that projections from current to future A-T production suffer from both *ceteris paribus* and automatic *mutatis mutandis* fallacies. The first fallacy Myrdal (1968: 1946–49) suggests is that of assuming that the growth/spread of A-T production would occur with "other things being equal." But other things will *not* stay constant as A-T production expands. For the first two classes of avocational and quasi-vocational workers, alternative opportunities for work benefits will diminish, as A-T drives out conventional industrial production: both technically challenging and materially rewarding "inappropriate technology" opportunities will shrink. Thus, A-T workers will have to accept A-T work conditions as *permanent vocational* opportunities. How will these workers respond to A-T limits under such conditions?[7] It is unlikely that their views of A-T opportunities will remain static.

The second Myrdal fallacy, automatic *mutatis mutandis* (1968: 1946–48), presupposes that no other changes will occur as a result of A-T diffusion, to resist this transformation. Among the many socioeconomic ramifications of A-T diffusion would be a change in the capacity of the state to provide a wide range of social welfare goods and services, including some public or collective goods now available in advanced industrial societies (Schnaiberg, 1982b). If capitalization of production is substantially reduced, so will tax revenues and associated state expenditures be reduced. To offset this loss, workers must: (a) forego such goods and services, (b) submit to wage reductions to provide tax revenue, (c) increase labor productivity through longer work hours or harder labor to generate sufficient surplus, (d) increase labor productivity to generate sufficient surplus to allocate some of this to capital, which will then be used to enhance capital-labor productivity and allow tax revenues to rise, or (e) perform volunteer labor to provide these state/public goods and services (Schnaiberg, 1980: ch. 5). In other words, all of these options suggest increases in worker obligations over the present A-T production organization, which currently coexists with and is subsidized by taxes from "in–A-T" (inappropriate technology) (Rothschild-Whitt, 1979). Again, then, worker conditions cannot simply be projected from current A-T organization.

These three critiques suggest a rather more discomfiting view of current A-T production. Contemporary A-T production is first and foremost a dissenting social movement product: while production and its technology appear to be its "product," social dissent and social mobilization of dissent are in fact the primary product of contemporary A-T. Equity proposals, not efficient production operations, are the main outcomes. These actual "products" currently depend on the surplus generated by the industrial system's "treadmill of production" (Schnaiberg, 1980: ch. 5). The social wages of workers in A-T—and often even the money wages for the paid workers—thus depend on the continuation of the treadmill's production. But the paradox is that the movement's goals will eliminate the treadmill, if the movement is successful. And this elimination will change the form and content of A-T work, including both social wages and fiscal wages. That is, routinization of A-T will lead to its transformation. As A-T becomes a dominant, routine form of production, the social wages associated with dissent will shrink, and the offsetting material supports outside A-T will diminish and vanish. Moreover, A-T workers will presumably have to perform production functions, including many "dirty" jobs currently assigned to the treadmill of production.

What does all this suggest about projections from current A-T production? In Schumpeter's (1947: 200–5) language, the only way A-T work could be fully diffused throughout the economy would be if economic leaders were "demigods," able to motivate all workers with wisdom and ideological animation and/or precisely the right balance of ideological and material incentives. In addition, all workers would have to be "archangels," internalizing the values of A-T (peace and permanence) completely and in perpetuity. Neither of these conditions is likely to hold in any system, as Schumpeter has so graphically argued. Therefore, the A-T movement needs to consider a wide range of worker traits and necessary socioeconomic incentives to respond to the reality of worker needs and demands in the future (Rodgers, 1978). To do so, leadership of the A-T movement must transform its sense of the movement away from idealism and into a clearer sense of the dilemmas of social production under A-T. These dilemmas are spelled out in the following sections, as sketches of the kinds of issues that routinized A-T production would have to confront.

POLITICAL GOALS OF A-T: MACROSTRUCTURAL DILEMMAS

One of the conclusions from the previous section is that A-T might have problems of political legitimacy among "workers" because of the socioeconomic transformation A-T will bring to the current treadmill of production. Among the necessary transformations will be a reduction in total surplus generated by production. This is so because, regardless of the social and ecological costs of the treadmill, there is simply no question about its being the most efficient social production system in generating social surplus (Schnaiberg, 1980: ch. 5). If A-T is successful, it will be because it has convinced politicians that present or future socioenvironmental costs are greater than the benefits of this expanded surplus. At present, organized labor (Hannon, 1975) has little commitment to A-T, for example, because it sees the benefits of the treadmill and the costs of A-T: social science can document many of these benefits, as well as some of the costs (Mishan, 1967, 1973; Thurow, 1980; Lave and Seskin, 1970; Barkley and Seckler, 1972). To date, the A-T movement has apparently made few inroads in such a constituency, preferring to deal with movement participants and poverty groups (Penner, 1978; Pitts, 1981) as main constituents. But some shift will be necessary to indicate to much broader strata of organized labor and the middle class why A-T is in their work-related interests. How this will be done is unclear: what is clear is that to do so many A-T ideologies will have to be transformed, in order to earn political legitimacy, much as the present treadmill has traded off greater welfare or equity investments to sustain the political legitimacy of high-efficiency treadmills (O'Connor, 1973; Schnaiberg, 1980: ch. 5). In the case of A-T, this will mean foregoing some equity and/or efficiency ideals of A-T, since workers are not archangels.

There are parallel problems among production or class elites in the United States. The challenge of A-T to the treadmill has not been clearly articulated, but to achieve the stated goals of the A-T program will require at least the temporary displacement of present economic elites (Gamson, 1975, 1980; cf. Goldstone, 1980; Schnaiberg, 1982b).

These "elite workers"—managerial and ownership groups—wield substantial power under the treadmill of production and have no motivation to abandon their positions. To date, the A-T movement has largely ignored these classes, preferring to argue about technology rather than about the reorganization of current production organizations (Schnaiberg, 1982b). But it is likely that the talents of many current managers will be necessary to administer some realistic form of modified A-T (or modified treadmill) production in the future. A-T "demigods" simply do not exist in sufficient quantity to reorganize total production systems. Again, inducements and controls will be required to achieve both the political compliance of current elites and their participation in A-T reorganization of production. This will likely entail some "inappropriateness" in A-T production, and/or some "hardness" in soft energy paths (Rossin, 1980).

At present, therefore, we only have a shell of a necessary A-T movement for work reorganization. It is as if A-T had no displacement goals (Gamson, 1975; Schnaiberg, 1980; ch. 8)—that it was constructing a new production system on the economist's proverbial featureless plain! Only in this way could the essentially non-conflictal nature of A-T be maintained: the current system of A-T is a paradigm of order, then, when it must itself be transformed into a paradigm of conflict (Buttel and Flinn, 1975; Buttel, 1976). Conflict will center on the goals and means of altering both the levels of surplus and the allocation of surplus under alternatives to the current treadmill of production (Schnaiberg, 1980: 205–209). This is, after all, the same type of conflict that underlies the current treadmill as well (Schnaiberg, 1980: ch. 5; O'Connor, 1973). But for this conflict to be orchestrated, both proponents of A-T and the social science observers of the movement need a far clearer sense of the likely dimensions of conflict. These directions of conflict grow, in part, out of the goals and proposed means of the movement and require an analysis of the most striking features of A-T movement goals, juxtaposed against the rather limited means they have proposed for reorganizing work.

THE CHALLENGE OF A-T VERSUS ITS MOVEMENT PREDECESSORS

As noted earlier, most American social movements have had redistributive goals: they have articulated demands for new rights (Gamson,

1975; Goldstone, 1980). Similarly the A-T movement emphasizes rights; however, reorganization of productive work entails obligations as well. To achieve its goals of peace and permanence, A-T production must entertain concerns about efficiency as well as equity (Okun, 1975). It must perforce be concerned about the level of societal surplus if social needs, let alone social demands, are to be met under A-T (Thurow, 1980). More directly and immediately, substantial social constituencies must believe these needs will be met before A-T movements can achieve significant legitimacy and power.

But we now confront a serious analytic dilemma. No prior social movement has ever attempted to incorporate equity and efficiency goals for the entire society.[8] How do our analyses of prior social movements help us understand or predict the future trajectory of A-T, then? We can only point out some of the relevant differences, and suggest the necessity for movement recognition of these as significant departures from social movement history.

Most social movements have had some "political" targets in that they have sought to redistribute some social surplus, some rights from work or the products of work. Both the civil rights and anti-poverty movements sought to incorporate the unemployed and underemployed into the benefits of both wage labor and/or transfer payments from the state (which were supported by labor and capital productivity). Union organizing movements aimed at better pay, shorter hours, improved on-the-job health conditions, and other prerequisites of labor (for example, Rodgers, 1978: ch. 6). While many of these movements have undoubtedly had consequences including new worker obligations, such obligations were not a component of the demands that social movements placed on their social problems claims (Spector and Kitsuse, 1977). Even radical political movements—extending to revolutionary forms—rarely placed explicit demands for particular forms of work on constituents. Such demands were often bureaucratically negotiated after the success of the political movement.

But the A-T movement is calling for a *change in and of work.* While its ideology highlights the necessity for change and the benefits of change, it is simultaneously raising worker consciousness about the potential costs of work change as well. Non-responsiveness or resistance by organized and unorganized labor (as well as investors and managers) has been interpreted by A-T proponents as simply due to a lack of consciousness-raising, thus leading to ever more suggestions for information dissemination, demonstration projects, and so forth. An alternative view is that a great deal of resistance is due to preferences for current work conditions versus those hypothesized under A-T (among those who are knowledgeable about A-T). And as more workers become exposed to A-T ideas, it may be suggested that a substantial proportion will also reject A-T work. In short, A-T requires a shift in worker motivation but entirely lacks a theory or agenda for worker incentives to change. In part it does so because of the one-sided view of work as benefit, rather than the duality of work as obligation-right. This is a domain in which managers of treadmill production are steeped, as part of their mission of maintaining and extending worker commitment and productivity. With A-T dismissal or ignorance of work-as-obligation, neither commitment nor productivity—nor political support of A-T—is likely to rise.

One of the simpler ways of expressing the differences between A-T and its movement precursors is that A-T anticipates a normative commitment to work (Rothschild-Whitt, 1979) rather than a managerial expectation of utilitarian motivations (Etzioni, 1964: 59–61). Past social movements have indeed had normative commitments from their participants, but these have been avocational commitments from such participants (as I noted for current A-T participants as well). In terms of their vocational commitments, social movement participants have not necessarily had any normative commitments. (Indeed, that is often why their participation in social movements is so important to them.) Movement work has thus been normative, and occupational work has been primarily utilitarian. The only exception to this—communitarian movements—only proves the rule.

Communitarian movements, ranging from the sects of nineteenth- and twentieth-century America (Baker and Hertz, 1980a, b) to the contemporary Israeli kibbutz, have largely been minoritarian or self-consciously deviant movements.[9] They have, in Robert K. Merton's terms (1957: chs. 4–5), been retreatist when confronted with the utilitarianism of modern industrial organization, choosing to withdraw and form their own production organization as a total institution. While some proselytization has often gone on, virtually all of these sects have accepted the need to remain sects and not become society-wide normative churches. Insofar as they have been successful in their withdrawal, it is often because they have accepted limited incursions *into* broader society, as the tradeoff for limited intrusion *of* the broader society. Even with this compromise, few of these sects have had long lives because of the contradictions between their normative organization and the utilitarian norms of the broader society (Hawkins, 1975).

What this exception suggests is the fragility of a movement with strong normative expectations (that is, a movement of "archangels," in Schumpeter's terms), because of competition from dominant social values and behavioral expectations in the broader society. These historical examples indicate the relative durability of the latter social norms of utilitarianism in work organizations, in sharp contrast to this fragility. This suggests the difficulty of the task confronting A-T, which seeks a rebellious rather than a retreatist solution to the incompatibility of utilitarian expectations of workers and the need for peace and permanence in production organization (cf. Rothschild-Whitt, 1979). This utilitarian model is contrasted with the A-T expectations in the next section, and the microstructural contrasts in typical work situations in the subsequent section.

A-T ORGANIZATIONAL MODELS VERSUS MANAGERIAL MODELS OF EFFICIENCY

If A-T models are qualitatively different in their goals and means from social movement precursors, they also differ from contemporary management institutions. If A-T leadership is to assume the role of "demigods" of production reorganization while archangels are not available to fill the roles of workers in A-T organizations, then the A-T leadership will have to address many of the problems of contemporary industrial management (Rodgers, 1978).

Within contemporary production organizations, the social (or class) relations between owners/managers and workers are primarily utilitarian, as noted above. That is, workers are kept as long as they are sufficiently productive for a profitable operation (socially or privately profitable, in socialist versus capitalist ideals). Similarly, workers stay as long as social wages in the enterprise are either adequate or competitive with alternative opportunities. Contrary to neoclassical economic analysis, however, this is not simply a "technical" system but a "political" one as well (O'Connor, 1973), as individual owners/managers and classes or sectors of management exert "political" influence over classes or sectors of workers. Opposition to or support for labor organizations, opposition or support for social security provisions and related activities in effect eventually alter social relations between owners/managers and workers by influencing power and resource differentials in the struggle for shares of production surplus/profits (Rodgers, 1978; Schnaiberg, 1980: ch. 5). But the principal task of managers is to coordinate production processes and worker demands for social and fiscal wages to ensure that sufficient profits remain for the enterprise to remain in the competitive marketplace; parallels exist under state capitalism or socialism as well (for example, Lindblom, 1977). In large part, this centers on the efficiency of production, ensuring that both labor demands and profitability can be met, by the adjustment of production technology and labor needs, and so on.

While worker demands must be met, managers need not concern themselves simultaneously with social equalities or with worker needs, except insofar as workers' threats to withdraw crucial labor power translate such needs into effective sociopolitical demands within the workplace. In contrast, A-T movement leaders imply that A-T "demigods" can and will relate production technology and the relations of production to workers' needs (Rothschild-Whitt, 1979; Penner, 1978). The concept that workers will have to organize to make demands against A-T managers has simply not arisen in the movement literature (cf. Schnaiberg, 1982b, c).

Such an A-T model of workplaces suggests that the workplace will become a microcosm of future society, in which equity and efficiency will be simultaneous concerns inside the workplace. In contrast, the contemporary division of labor in modern industrial societies is between the goal of workplace efficiency (a technical objective) and social distribution (a political objective, for example, Friedland, et al., 1977) aimed at approaches to equity. Within the modern treadmill of production, efficiency and equity are unlikely to coexist as goals of a single organization; government and social movement organizations tend to deal with equity issues, while the private production sector focuses on efficiency issues. There are many points of interaction between the two, as when industrial advisory groups help set government tax policies which redirect social surplus into either efficiency or equity allocations (Schnaiberg, 1980: chs. 1, 5, 6). Similarly environmental protection enforcement affects conditions of efficiency within the production workplace, in the name of equity entitlements to a clean environment (for example, Tucker, 1981; Rosencranz, 1981).

With these important exceptions and some additional ones in public sector organizations that are aimed at "community development" in one way or another—such as the Tennessee Valley Authority (Selznick, 1966) or the Housing and Urban Development Department—there is relatively sharp differentiation of major tasks of production and welfare organizations. This will be blurred in the administrative structure of A-T in the hypothesized future A-T society, since somehow equity is to be one major component of production organizations (Rothschild-Whitt, 1979). How equity and efficiency will actually be merged at the workplace level is not clear (cf. Okun, 1975; Thurow, 1980). But it implies some substantial differences between current managers of efficiency in the modern treadmill production organizations and the future A-T managers in the non-treadmill production organizations. At the extreme, the future A-T society will have to find some ways of approximating a "demigod"-like administrator to replace the narrowness of current administrators of production. This suggest that current managers will anticipate their being displaced, and thus they and their owners-investors will resist A-T (Schnaiberg, 1982; Goldstone, 1980; cf. Gamson, 1975, 1980). Implementing this dual goal structure at the workplace level will pose enormous problems as well, as illustrated in the following discussion.

THE MICROSTRUCTURE OF WORK UNDER A-T

An ideal-typical A-T production operation is likelier to be something approximating a batch versus continuous-process technology (Woodward, 1965; Blauner, 1964), because of the resource-conserving aspects of this technology. In order to employ labor rather than capital, many elements of production currently in use would have worker controls substituting for mechanical or electronic controls. This is, after all, what "accessible technology" implies at the workplace level. One of the concomitants of this deautomation of production is the increased watchfulness required of workers: constant movement and/or scrutiny of processes is likely to be required, to ensure that production goes smoothly and that products are of high quality.

The positive side of this is a renewal of a quality we may call "craftsmanship," though this may be too generous a term. It is not that each worker will be responsible for the production of an entire product—the costs of products under such a system would be far too high under any reasonable (equitable) wage conditions. Unless craftspeople were like indentured servants or unpaid apprentices, production costs would be too high for this to happen. Thus, something between pre-industrial eotechnics and modern neotechics (Mumford, 1973) would be required, in which modest levels of capital would be combined with watchful/attentive workers to produce socially usable products: a "quasi-mass production" or "small factory" model of production.

Such types of work would have a range of demands on worker energy that differs from many current occupational roles under the treadmill of production. For most solid skilled working-class workers (Levison, 1974), it likely entails an increase in physical labor and probably an increase in emotional-mental labor as well (for example, making decisions and worrying about operations more than in automated or semi-automated factories in which attentiveness may be lower). For unskilled lower-class workers, it may mean no more physical labor, but substantial increases in emotional-mental labor. For technical-clerical types of workers, there is likely to be an increase in physical labor and perhaps some increase (or decrease) in emotional-mental labor. Finally, for highly skilled professional-technical workers (including managers and intellectuals in the academy and elsewhere), there is bound to be increased physical labor and perhaps a decrease in some emotional-mental labor (associated with the risks and uncertainties of professional production, say), with some increases in other emotional-mental labor (boredom, interruptions of tasks, and lack of concentration on a single goal).

These are, of course, speculative discussions, since we do not know what A-T production would actually be like. But recent experiences on a limited scale (Pitts, 1981)—such as neighborhood greenhouse systems in poorer neighborhoods—require us to think about such issues, in two ways. First, what work obligations will actually exist under A-T production? Second, how will these obligations compare with those faced by different groups of workers currently in our social production system—including the unemployed or underemployed, in whose name A-T proponents or soft energy proponents (such as Hannon, 1975) proclaim the social value of A-T production reorganization most loudly? Both of these are, then, necessary components of a true comparison of A-T production with treadmill production (as in Morrison and Lodwick, 1981), in terms of work and worker attitudes to the changes proposed. They involve a detailed working out of what "accessible technology" actually will mean, as it begins to take over all social production tasks (Penner, 1978; de Moll, 1978).

What the speculations above suggest is that "soft energy" will often require "hard labor" in order to operate with even modest efficiency. James P. Pitts (1981) indicates, as did Lane de Moll (1978), some of the problems of integrating educated professionals and unskilled poor in a single project or organization. Pitts's work traces the history of a neighborhood technology action group and their attempt to implement self-supporting greenhouses as a source of urban food supply, or at least for production of green plants to be used for cash income to help offset poverty conditions. The realities of both construction and greenhouse operation—ranging from the difficulties of skill transference to worker motivation to regularity of volunteer commitment—permeate the operating workforce behavior (the "archangels"). Equal frustration existed at the project leadership level (the "demigod" level), because of the problems of mobilizing labor power and dealing with the constraints on production imposed by city agencies (zoning and so on). The applicability of these quasi-vocational experiences (a mixture of vocational and volunteer activities on the part of both project leaders and project workers) to future routinized A-T work is limited, as noted earlier. However, these limited illustrations of problems in worker motivation (leaders and workers) indicate a range of social control issues that will have to be addressed if A-T production is to have sustained efficiency (cf. Rothschild-Whitt, 1979). Equity—or equitable allocation of social surplus under A-T—is meaningful only where there is sufficient production to generate surplus (Schnaiberg, 1980: ch. 5, 9). If A-T production falls short of social needs, then it is equity in allocating scarcity that is all that A-T proponents can hope for—and that raises many problems of maintaining political legitimacy for an adequate economic system.

The limited empirical evidence we have for the actual types of A-T work indicates that many of the traditional problems of labor management-recruitment, training, supervision, discipline, and material rewards—will also be problems that need to be addressed by A-T leadership (Rothschild-Whitt, 1979). As de Moll (1978) indicates, hard work is involved in attempting to be self-reliant. Educated workers may idealize their leadership roles in this struggle for self-reliance, but their own work conditions are very likely to change substantially under A-

T production. Far less distance is going to exist between direct production workers and managers, and far more boring, mundane, "dirty work" is likely to be demanded of the white collar workers staffing A-T production (Rothschild-Whitt, 1979). The alternatives include increasing the scale and/or efficiency of production to free white collar workers to be creative thinkers rather than production tinkerers, interacting with workers and physical technology on a daily or even hourly basis. But this will require changes in the scale and probably the physical and social technology, and in the social relations of production under A-T. From the vantage point of these workers, this major modification of the "smallness" and "accessibility" of production technology may be well worth it. But the costs will be a weakening of A-T goals, an incorporation of some treadmill-like relations and forces of production (Rothschild-Whitt, 1979).

From the vantage of many workers, the need for sustained, physically demanding work that is a likely component of much of A-T production may not be acceptable for an indefinite future. Workers may accept this set of conditions as temporary until energy or other resource restraints are dealt with, but over the long run, they are likely to demand adjustments in the surplus generation system to permit greater freedom in and from work. Some increased efficiency within resource constraints may be possible and likely, if workers welcome this efficiency increase, knowing that there will be an equitable sharing of the benefits. But there is no automatic limit to this demand, and some form of continuing socialization and/or social control will be necessary to prevent demands for reintroduction of treadmill-like increases in economic efficiency. This also indicates roots of resistance to acceptance of A-T workers (Schnaiberg, 1982b).

Recent history suggests other empirical illustrations to reinforce the cautions above. The post-1948 production history in China was one with dual concerns of efficiency and equity, given Maoist principles of politics and economics. For a substantial period of time, labor-intensity and presumably highly equitable allocation of surplus coexisted, culminating in the Cultural Revolution and associated production models of "backyard production" (for example, Anderson, 1976; cf. Parish and Whyte, 1978). In recent years, however, we have observed two kinds of changes. First, more data have emerged on the deviations from an ideal balancing of efficiency and equity goals/means in China in the 1948–78 period. The picture we now have is one of continuing conflict among the leadership of "demigods" and disruption among the urban and rural working "archangels." Rather than a smooth adjustment to these dual goals, the actual history now resembles a dialectical conflict, with much zig-zagging, because of the strength of equity versus efficiency forces within China. The same tensions are likely to accompany any shifts toward A-T in the United States, for some of the same structural reasons; although absolute deprivation is much lower in the United States than in China (that is, surplus is much higher), relative deprivation impulses are equally strong and enduring. Second, in our own society we can point to the decline of the counterculture and its rejection of materialism (cf. Ingelhart, 1977),[10] and the rise of "ambition" again in the young educated cohorts as a caution analogous to that suggested by the recent shift toward "modernization" in China. Both modernization in China and ambition in America are associated with the same process: making more surplus available. In the absence of a truly imaginative production scheme, in general this implies more shifts toward treadmill-like production (Schnaiberg, 1980: ch. 1, 5), including both more resource consumption and a more hierarchical form of production organization, as well as more capital intensification.

Thus, the routinization of production under A-T rules will have to content with two kinds of ongoing resistances: (1) the work will be either too hard or too long; and/or (2) the rewards *from* work (or *in* work) will be too little. Given the values of equity and democracy proclaimed by A-T proponents, moreover, it will be hard to dismiss such protests "for the good of the nation," as the contemporary Reagan administration can do with the currently deprived workers in America. Unlike current theorizing, democracy is not merely an end in itself, but one means of articulating discontent and resistance (Rothschild-Whitt, 1979): how then will A-T structures permit a responsiveness to these discontents with work as its rewards? Will A-T demigods increase rewards—or will they stifle dissent? Increased rewards will demand compromises with A-T ideals of non-treadmill technologies of production—and stifled dissent with A-T ideals about democracy in production. Once more, decisions about the forces and relations of production will have to be negotiated under the same kinds of political difficulties that currently bedevil contemporary treadmills of production: to believe otherwise is to substitute ideology for analysis.

Finally, this discussion would be incomplete without a sense of the non-universality of work even under A-T. Not everyone in the society will be workers—there will be children, child-care parents, the sick, aged, and handicapped. To provide for their maintenance, workers will have to forego some of their own personal rewards from work—or work technology will have to undergo a sufficient increase in efficiency to cover all the non-work "reproduction" tasks of a society (Schnaiberg, 1982b). The politics of welfare systems under A-T may resemble those in the contemporary advanced industrial societies more than some utopian ideal, with continuing conflict between workers' shares of surplus and non-workers' shares. Again, both production and welfare politics will have to deal with this (Schnaiberg, 1981), as within the current treadmill system.

CONCLUSIONS

From the above discussions, the beginnings of a sense of the dialectics of A-T production emerge, rather than the sense of a linear transformation of society that will resolve all underlying conflict in and about production (Schumpeter, 1947; Schnaiberg, 1982b). In short, "the problems of production" will not be permanently and peacefully solved under A-T, any more than they will be solved under the treadmill (Schnaiberg, 1975; 1980: ch. 9). This suggests the following observations:

1. The future will see a struggle between equity and efficiency interests, as the past has seen.
2. There will be substantial resistance to the transformation of the treadmill of production into A-T production, because of the large surplus of the treadmill, and its allocation to a variety of interests.
3. Additional resistance to this transformation will arise because of the changes of work itself that A-T will entail, and the undesirability of many of these changes to workers.
4. Even after "transformation is complete," the resistances noted above will reemerge.
5. Perpetual struggle and resistance against drift back to the treadmill will be likely after the successful transformation to A-T: to deal with this, substantial social science insights are necessary to complement the technological insights currently dominating A-T—for example, problems related to worker motivation, political organization, socialization, social control, and so on.

Our role as social scientists is to raise these issues, even if A-T partisans are disinterested in them. If we support A-T, we must articulate its problems and the social changes necessary to reduce them; if we oppose it, then we must articulate at least the costs of A-T in an empirically grounded way.

NOTES

1. In this regard, this effectively reinforces nineteenth-century American attitudes about the nobility of labor (Rodgers, 1978), attitudes that became increasingly unsupportable as conditions of work deteriorated from a craft ideal (for example, Braverman, 1974).

2. Some discussion of this kind of organization exists in Rothschild-Whitt's (1979) analysis of "alternative" institutions, in which motivational problems are also considered. She writes in support of these institutions, many of which have limited market production aspects but provide services. That is, in some ways her sample of organizations is more favorable to A-T-like conditions than will be the more routinized production of physical goods (cf. Baker and Hertz, 1980).

3. Rodgers (1978) has a rather focused discussion of worker attitudes in nineteenth- and early twentieth-century America. In general, he traces a decline in the nobility-of-work ethic and the rise of movements, initially to prevent work fragmentation and mechanization, and later to limit work hours and raise wages when mechanization increased. This seems to support A-T arguments and refute my own. However, the level of craft ideals which these nineteenth-century workers sought was below the level of mechanization of A-T technology (Schumacher, 1973; Lovins, 1977). Thus, if my interpretation of A-T is accurate, workers would find even this level of technology routinized and deadening, as opposed to idealized crafts. Indeed, there is, as Rodgers notes, much dispute about the social desirability of cottage industry production which predated the factory system.

4. Rothschild-Whitt's (1979) sample, as noted above, does not represent the goods-production sectors well: her only "production" example was a newspaper one. This is hardly representative of most routinized factories of the treadmill of production, for which A-T must find substitutes. Hawkins (1975) analyzes conflicts in four "retreatist" communes, which withdrew from the mainstream to achieve communal production-distribution goals. Unlike Rothschild-Whitt, his emphasis was on the contradictions between efficiency and democratic-participatory equity and the failure of these communal organizations to thrive.

5. Rothschild-Whitt notes, in passing, the throughput of staff of alternative institutions; they use these organizations as waystations on a path to employ-

ment in treadmill organizations. But she gives us no data on turnover or on staff motivational features.

6. Both of these authors talk about wages, but in neither analysis is it clear how wage-supply relates to workers' needs—that is, whether the motivation is long-term occupations, short-term occupations, or some other mixture of relations to the organization (de Moll, 1978). It seems to be in bad taste to discuss workers' standards of living, for few accounts are given of this.

7. Again, two of the richer accounts of A-T-like organizations—Rothschild-Whitt (1979) and Pitts (1981)—give us little sense of how these organizational experiences fit into the life-cycle expectations of leaders/workers (though Pitts offers some data on leadership). My thesis is that many people can commit themselves to A-T work now only because if their ambition rises or family needs change, they can slip back into the treadmill organization—much as many of the counterculture participants shifted.

8. The counterculture (for example, Hawkins, 1975; Rothschild-Whitt, 1979) in the United States came closest to this ideology. But it fell short of the A-T model by (1) becoming retreatist, with isolated communes rather than proselytizing community institutions (Hawkins, 1975); (2) making accommodations with the treadmill by forming special-function organizations (Rothschild-Whitt, 1979), and interacting with and accepting much of the treadmill structure for their economic sustenance, in order to survive; and (3) holding far more to equity than to resource-efficiency goals.

9. The kibbutz movement has always been a minoritarian, deviant one within Palestine and later the State of Israel (for example, Baker and Hertz, 1981a, b). Two features of its equity-within-efficiency dialectic which have been studied are (1) the relations of inequality between men and women (Baker and Hertz, 1981a, b; Blumberg, 1974, 1977; Bowes, 1978; Paden-Eisenstark, 1973; Rabin, 1970; Schlesinger, 1977; Talmon, 1972; Tiger and Shepher, 1975) and (2) the relations between communal members' labor and the hiring of paid outside labor (Rayman, 1977; Baker and Hertz, 1980b). For both of these, compromises with ideals have proven necessary over the long run.

10. While Inglehart (1977) stresses that "materialism" is a lesser value to "post-materialist" respondents than justice/democracy, he also cautions that post-materialists are likely to be taking materialist gains for granted. Thus, recessions, depressions, or the A-T suggestions for reduced consumption may lead to resistance among post-materialists if they reduce current levels of consumption—as A-T seems likely to do. Hence, "post"-materialism can be reinterpreted to mean "support for *no-growth*" rather than "support for *attenuated* consumption"; the latter seems more realistic for many A-T proposals, though few A-T proponents dwell on this.

17

Counterpoint: Use Where Appropriate: The Scientific Appropriateness of Appropriate Technology

HARVEY BUCKMASTER

I am pleased and honored as a physicist and engineer to have the opportunity to participate in this colloquium and to provide commentary on Professor Schnaiberg's paper. It would be presumptuous of me to attempt to discuss Schnaiberg's paper from a sociological perspective, but my views are strongly affected by social constraints since I am interested in the extent to which science can be a humanizing activity. I see society in terms of the creation and extinction of its mythologies. Consequently, it seems appropriate to raise my philosophical underpinnings as a pure and applied scientist.

My interest in the impact of technology on society is both longstanding and pragmatic. I have concluded that many of our social and environmental problems have arisen because we have accepted as fact, without critical assessment, the Madison Avenue charge that "what is technologically feasible is also socially desirable." The consequence for the developed nations is that they have experienced extremely significant social and environmental as well as economic transformation during the past two centuries. During the nineteenth century, the roles of the manufacturing and trading companies were one and the same as the national interests as articulated in the colonial policies of Great Britain and later Germany, France, the Netherlands, and the like. The building of the railways in Canada and the United States is an excellent example of how the interests of the private and public sectors can become intertwined.

The twentieth century has seen additions to this group, including the United States, Japan, and the USSR. It has also seen the rise of a new form of economic imperialism, the multinational corporation (except in the USSR and China), which filled the vacuum created by the demise of territorial colonialism. We are now in danger of accepting a new mythology that "what is in the best interests of these conglomerates is in the best interests of the sovereign powers in which they operate." Canadians and other marginal countries have more reason to distrust this new mythology than most developed nations.

It is against this background that I approach the concept of A-T. It is my view that A-T has a legitimate scientific basis and that it should not be perceived as a rejection of technology. My interest arose because of my concerns about the environmental impact of increasing energy consumption. These concerns have become more acute since I have discovered that a growing number of pure and applied scientists would create a new mythology that only those areas of science leading to technological innovation which is intellectually challenging should be supported and realized. It is paradoxical that this attitude is diametrically opposed to that espoused by those proponents of A-T who are not anti-technology. It is important to observe that A-T is a realization of the application of the guiding principle known as Occam's razor. The simplest solution to a technical problem is frequently that which is the most reliable and economically viable. Moreover, this elitist attitude of some scientists and engineers is subject to at least indirect public dissent by the opponents of nuclear energy and other high-technology projects.

The concept of A-T is not new, but it was revived with renewed vigor by E. F. Schumacher and less explicitly by Amory Lovins. Schumacher used this concept to deal with the complex interface of Western technology with the Third World. Schumacher recognized that culture, in the broadest sense of the word, encompassed technology and that attempts to introduce alien technologies could not be successful unless the social group was receptive to cultural change or adaptation. An analogy is the existence of abnormal cells in the human body. They are normally destroyed by the antibody immunity system. However, some abnormal cells are, or become, resistant to this system and survive and reproduce. This is the origin of cancer. Schumacher realized that Western technology could play an effective role in liberating the Third World from many forms of crippling labor through its introduction in an "appropriate" form. He recognized that a humane social and economic revolution was possible if "appropriate" technologies could be introduced throughout the underdeveloped world. He founded the Institute of Intermediate Technology in order to develop the appropriate technologies. It is my assessment that Schnaiberg has used a definition of "appropriate" technology that differs significantly from that of Schumacher and many other followers of the A-T movement.

One has a strong feeling of *deja vu* as one reads Schnaiberg's paper. One is reminded of the Fabian Movement. Schnaiberg's assessment of the motivation for the transformation to an A-T society does not differ from the Fabian views. It appears that everything Schnaiberg has written could have been said about the proposals of the Fabians and indeed, probably was said by their critics. The analogy is compelling since the Fabian Movement was a direct result of the intellectual disenchantment of the Victorians with the society that had resulted from the nineteenth-century form of the Industrial Revolution. They extolled the virtues of craftsmanship and of hand-made goods. John Ruskin and William Morris were two of the apostles of this movement; they capitalized on a yearning for the mythic "good old days." It is paradoxical that these groups of intelligentsia and their followers have never known what it was like to be a laborer in those "good old days." This is the area where Schnaiberg is most effective. His arguments are based on the absence of altruism in the laboring class that has been frequently romanticized by such intellectual movements.

It is perhaps appropriate to recall the perceptive comments of Marx on this subject. One finds a similar pining for the past among some of the followers of A-T.

I am convinced that it is impossible to provide a unique definition of A-T. Many advocates in developed countries, including Lovins, do not reject technology. Instead, they argue for the best compromise between efficiency and reliability that is environmentally as benign as possible, since they are convinced that current trends toward larger scale technologies promote inefficiency and decreased reliability. In short, technology has outgrown itself. This is the basis of Lovin's ideas and the origin of his "soft energy paths." It is, perhaps, unfortunate that this approach has provided a convenient niche for the anti-nuclear lobby. This group has raised another set of concerns and issues that have detracted from and confused the basic thrust of the A-T movement. Lovins himself is responsible for this confusion since he is strongly anti-nuclear. Readers can detect a romantic component in Lovins's conviction that humankind yearns for a simple technical formulation of the way to achieve a high quality life. I agree with Schnaiberg, since quality of life is always much closer to survival for the laboring class than it is for middle-class intellectuals.

It is perhaps useful to note that many supporters of the A-T movement use it as a proxy for a more profound form of societal dissent. It is easier to be anti-technology than it is to fight the increasing centralization and bureaucratization of our systems of government. Many citizens feel alienated since it is impossible for them to understand, let alone make informed decisions about, contemporary political issues. The swing to the political right is a manifestation of this frustration. However, it will not solve the underlying problem and represents but another version of the yearning for "the good old days" when everything was simple and issues were perceived in black and white terms. This yearning appears to be primarily a North American phenomenon and is probably due to the strong pioneer mentality that remains after two or three generations of agricultural development. It is a significant part of our romanticizing and mythmaking.

The A-T for intermediately developed portions of the globe must be less sophisticated because the degree of industrialization there is less complete than in the developed nations. Furthermore, the A-T is limited by the supply and transportation networks as well as by the general level of technical education. A reliable machine, say an electric water pump, may be of little use in an intermediately developed country where an electrical power grid exists but where its operation is unreliable. Consequently a wind-powered jack pump would be simpler to maintain as well as more reliable in fulfilling its role of providing fresh water for human and animal use. In underdeveloped countries, A-T takes an even simpler and less sophisticated form. It may not be feasible to introduce windmills to pump water if they are constructed with complicated metal parts that are difficult to repair or use complex gaskets that cannot be fabricated from, say, inner tubes from truck tires. Hence the design must be simpler so that the components can be fabricated from local materials when necessary. The design of such machines was the objective of Schumacher's Institute of Intermediate Technology.

The current transfer of the technology of developed countries to the underdeveloped portion of the world has been a very mixed blessing simply because it has been inappropriate. The consequence has been that the great majority of the people in these regions have continued to live in abject poverty despite the huge sums expended by developed countries on aid programs. It has had serious negative social effects since it has served to accelerate the concentration of wealth and power in the hands of a few.

It is tempting to speculate whether the A-T movement will spawn political parties as it emerges from its embryonic anti-establishment form of dissent. This presumes that the A-T movement will take this direction in developed countries. The alternative may be its absorption into the platform of an existing labor-oriented party. It should be recalled that the British Labour Party evolved from the Fabian Movement. Schnaiberg appears to have formulated the implication of an A-T society in a narrow way that might be appropriate to the United States which lacks a labor-oriented political party. Moreover, he assumes that the existing intensity of the conflict between blue collar employees and their employers will continue. The lack of agreed national objectives and social cohesiveness, as well as the undue emphasis on individual rights and freedoms unfettered by concomitant responsibilities, appears to characterize both the United States and Canada in particular but also most English-speaking countries. Japan is an example of a developed country that is avoiding many of these pitfalls. Its unique cultural traditions have enabled it to avoid most of these hazards and to successfully absorb an alien technology. It may be able to teach us many valuable lessons since both traditional and Western technologies are used, depending on the circumstances. Is this not an example of a culturally fine-tuned A-T?

Finally, as a pure and applied scientist, I want to emphasize that the scientific basis of A-T is very sound. The concept of matching is fundamental to that of efficiency. It is essential that we learn to utilize our non-renewable energy resources as efficiently as possible. A-T is the efficient approach to energy generation and use. It is for this reason that Lovins's "soft energy paths" have had significant influence on our thinking: thinking A-T is the way to unfetter ourselves from "the bigger and better mousetrap" mentality. Kenneth E. Boulding has given us the imagery of "spaceship Earth" to help emphasize that many of our resources are finite and non-renewable. This metaphor shifts the emphasis to renewable resources, particularly in the area of energy. It also reminds us that proper design and insulation are much more appropriate solutions than solar heating units and windmills.

We should be appreciative of Professor Schnaiberg's detailed examination of the social implications of one variety of the A-T movement. However, it would be a tragedy for humankind if his version turned out to be the only variety that is ever attempted.

VI
Colloquium Overview

18

Movements in Environmental Awareness

WILLIAM LEISS

I will not pretend that any overview can capture all of the details of what has preceded. I will only try to capture the spirit of the proceedings. It should not surprise you that there could be a number of perspectives that one could approach this subject with, and I would be equally interested in hearing yours in reply to this one.

What we have done here indicates that we are well within the third phase of the awareness of environmental issues. I think that everything presented here falls within such a third phase, and I would like to try to understand this third phase as a resolution of a tension that existed between the first two phases. I am in my deepest instincts incorrigibly a Hegelian. This portrayal of two phases and a third, which in a sense stands permanently between them, parallels to some extent the dichotomy of the political and the technical in Allan Schnaiberg's introductory paper but probably is not quite identical with it. In the first phase I see an emphasis on environmental philosophy, very broadly, conceived in the late 1960s and early 1970s and thereafter succeeded by a second phase, dominated by impact assessment, and a third phase which we are now in, of which this conference is a part. I do not know how to dub it. In any case, I would like you to keep in mind this idea of two poles—of environmental philosophy as the initial phase of the movement of environmental awareness, and impact assessment as the indication of the second phase. I think of them as not having disappeared, or having been subsumed by a third phase, but as still existing.

What we were doing here the last two days exists as a kind of tension between these two poles.

The first phase of environmental philosophy was dominated by the big questions. When I first went to the Faculty of Environmental Studies at York University and began a series of seminars with such interesting people as John Livingstone, we spent a lot of time on the basic questions, chiefly the question of the human relation to nature, as well as such questions as the rights of non-human nature, the rights of natural entities. Could there be legal rights for non-human nature? John and I went to a conference on that in Claremont, California, in 1974. The conference raised questions about wildlife conservation. Recently Livingstone released a book called *Fallacy of Wildlife Conservation*. We gave interdisciplinary workshops on eco-design, in which we talked about the need for a fundamental reorientation of lifestyles, which would be reflected in physical structures of living as well as interaction patterns. There was a time during this period of environmental philosophy when we began to think that environmental issues put us beyond the normal ideological controversies of capitalism and socialism. It was larger than that!

This phase was also characterized by a certain intuition that there were things of intrinsic evil—sometimes nuclear power is classified in this way. They were intrinsically bad, but not for any reasons that would be revealed in cost-benefit analysis. There was a time in which very dramatic slogans such as "small is beautiful" and "A-T" captured the imagination. If you missed this phase and want a more recent touchstone, there is a collection called *Ecological Consciousness* (published in Washington about 1980) which was the proceedings of a conference held at the University of Denver in the spring of 1979. This was a brave attempt, and one of the last, to bring forward these big questions, and to ask questions about the lifestyles of primitive peoples and their relation to nature and what bearing this had as a critique of industrial civilization. Some of us spent a long time on such questions. This phase has faded into the background. In summary, this phase could be characterized by claiming that "we can respond to environmental concerns only by a radical and total shift in our behavior patterns and in our modes of thought." The shift is based on our recognizing nature as an autonomous sphere vis-à-vis society; our programmatic sense could be summed up in the form of a warning: "Mother nature is trying to tell us something, we'd better listen."

This phase was quickly succeeded by a very different phase characterized by impact assessment. It differs in almost every detail from the big questions on environmental philosophy, even though it was related specifically to environmental issues in which certain things like risk-assessment came to the floor, especially in nuclear debate. Here it was suggested that we should proceed, not with reference to big questions, but by case-by-case analysis with pragmatic rules in which the single most important word was "tradeoff." It was characterized by the attempt to quantify intangibles. I have done a little study of one of these fascinating aspects—landscape assessment and the attempt to quantify the aesthetics of landscape perception. This came directly out of environmental impact assessment because the challenge was from the environmental movement as the first phase. The impact assessment advocates focused on economics, production costs, and so on. But there were intangible values that cannot be quantified. These were an obstacle to decision-making. The response from impact assessment advocates was that everything is measurable. They found ways of measuring and quantifying aesthetic variables under the rubric of benefit-cost analysis. This reflected an implicit priority in favor of economic development: development provided the emphasis for the impact assessment. However, this impact assessment/development coalition had already made certain concessions to the environmental movement, such as conceding the importance of externalities like the quality of life and quality of the environment beyond the factory gate. One can summarize this phase in a single image: the claim that we can respond to environmental issues by adding one additional variable to the decision-making matrix, "environment." Therefore, we do not have to change the decision-making process itself. And one slogan could sum up the mentality of this phase: "no problem, everything's still under control."

What we are experiencing is the result of the interpenetration of the influence from these two quite dissimilar first phases. We should look on the positive side of our environmental philosophy. It certainly did get us to think in a much more sophisticated fashion about things such as efficiency. Dr. Buckmaster mentioned the concept of matching technology to the actual needs. As a result of the environmental movement, a much more sophisticated concept of efficiency has been accepted. The movement also placed environmental issues permanently on decision-making agendas. And I think that also, in going beyond that, it raised some lasting sensitivity to concepts of intrinsic value beyond benefit-cost analysis, one example of which is the intrinsic value of wildlife habitat, not that this is a permanent achievement. Those I think are some of the positive features in the permanent contributions of environmental philosophies in its first phase.

This phase also had some negative aspects which quite properly resulted in its receding into the background until it could address these deficiencies in its formulation. First of all, such total shifts in behavior patterns and modes of thought as it said were necessary are simply out of the question. They cannot occur. And so if it is said "unless this happens, we die," then we are doomed. Such basic change is not going to be possible. Second, and even more seriously, the movement had another basic flaw. It was premised on the notion that there was an overriding issue, namely, "concern for the environment" that put into the background all the divisions of social interests. Yes, there were rich and poor countries as well as citizens. Yes, there were inequities between male and female, and racial divisions, and so on. But all of these must recede into the background according to the environmental movement, because of one overriding concern in which we all share, namely, the "liability of the ecosphere." This was an implicit premise, and this is false. Again, even if it were true, one could not realize it because those other divisions are too real in human experience to be put into the background.

On the other side, in the second phase, the impact assessment phase had positive dimensions. In its best form, it insisted on a serious concern for evidence in these debates. It was also correct in its insistence that we must recognize that whether or not we like it, tradeoffs will take place. We can pretend that they do not exist, but they will not go away; in fact they occur even when we deny them. It introduced a certain realism into the debate which was quite salutary. On the negative side of this second phase of impact assessment, there was a tendency that we all know, especially those of us who refrain from the vulgar statistical aspects of the social sciences and instead confine ourselves to the purer realms of theory. This was a fetish for numbers and for assigning dubious validity to numbers. Quite apart from the facetious remark just made, and in a serious vein, I still maintain that there is a lot of fiddling around with numbers in all forms of statistical social sciences. In its concern for extracting more from the environment and doing so more efficiently, in order to feed what Allan Schnaiberg has called the "treadmill of production," impact assessment cannot address the crucial question, "does more and more consumption really increase satisfaction and welfare?" This key question has been addressed subsequently, in the intervening period, in the literature on the social limits of growth.

Now when I talk about our perspectives here as avoiding these polarities although drawing on them in various ways and in a sense constituting a resolution of them, I use "resolution" in a sense that does not mean a "canceling-out" of these contradictory elements. Rather, we exist in an ongoing tension between these two early phases, and we draw our sustenance and our concerns from them in a variety of ways. This third phase, based on the tension between the first two, is the perspective that has motivated this conference in as far as its conclusions are concerned.

Let me describe the conclusions I have drawn from our discussion. My first conclusion is that *rationality is not determining*. The notion that science is an enterprise that can tell us what environmental standards are and should be, and thus what a responsible environmental impact should be, and following that, that the administrative-legal structure will implement these standards which we know to be true and correct, does not hold. It is an illusion. Both Thompson and Mazur and their commentators in different ways address this point. And it is addressed negatively in Nelkin's paper in the sense that one must face the fact that there is a widespread irreducible irrationality in the public that will not go away and that, as Allan Mazur has indicated time and again in this conference is not amenable to further facts. On the whole, rationality is not determining. By way of a second conclusion, *technology is not determining*. This matter is directly addressed in a number of papers. In the exchange between Peitchinis and Mills on the effects of the microelectronic processor revolution, it seems that the outcome of this development depends on the choices made in the wider social realm. It is not the technology that will determine the nature of those choices. This is also the theme of my exchange with Bill Reeves. And I heard at least some of that in the session with Schnaiberg and Buckmaster—A-T in and of itself has no rational content that is universally applicable. One must seriously qualify the meaning of appropriateness by specifying context. On the other hand I think Allan Schnaiberg should have conceded that if there are problems on that side there are a hell of a lot of problems on the side of the big boys and the massive industrial structures. If A-T has problems, then massive industrial structures would like to have only such problems. And negatively I heard in Carson's paper that it was the political economic imperatives, and not the technical imperatives, that govern development in such things as offshore oil.

So what is the determining factor?—well, very broadly, social relations as they bear on public policy. In this domain we find the grounds of the contradictory outcomes that appear in these wild shifts of public opinion to which Allan Mazur has called our attention. What are the internal contradictions in the public realm and the grounds of uncertainties in our thinking about modern technology that bear on, and are exhibited in, many discussions we have had? First, on the one hand there is a widespread commitment in the population, a passionate commitment, to industrial production in general and to its material benefits. On the other hand, we find determined opposition to particular negative aspects of production such as toxic waste disposal when they impinge directly on us: the famous "not in my backyard" syndrome. This is irrational from the point of view of the detached observer, but it is not going to go away by our calling it irrational. On the level of the private person it is a rational response, and on the social level it is irrational; we have to live with that contradiction. Second, I believe in the public as well: there is a general commitment to rationality in public decision-making processes, more so than in the past. Our society has made real progress in this sense. Yet, we find a seemingly irreducible irrationality on particular issues, for example, fluoridation and some aspects of the nuclear controversy. Mazur has called attention to the fact that no amount of additional fact will change a certain degree of public opinion. Hence, contradiction between a general commitment to rationality and some particular irrationalities also exists for us.

Finally, perhaps most generally, there is a contradiction in the public mind between, on the one hand, a general confidence in the power of modern technology to solve problems, and on the other, a nagging subconscious fear that our dependence on a complex technological apparatus which we really do not understand makes us vulnerable to unexpected catastrophes. These dichotomies or contradictions in the public mind between a rational and irrational component are what constrains decision-making and also affects the response to technological issues.

My conclusion is quite firmly that we should remain committed to a public inquiry process, which is well established in Canada. We should be so no matter how crude and imperfect an instrument it may be and no matter how much the latent irrationalities that I have referred to surface in these settings, as Allan Olmsted in the hearings in Warman, Saskatchewan, related they did. No matter how much that happens, we should remain committed to that process at least until we find a better instrument, and I think it will be a long time before we do. When we compare the Canadian model with that of other countries—and here I give my nationalist pitch—for example, with the French and German models, we find that a generalized lack of access by the public to decision-making forums leads directly to a permanent latent potential for violent confrontation. They do not have such public processes there. They do not have the ability to let off steam in these processes, and I think it is understandable that in such cases, you might want to fire a rocket when you simply cannot get through by any other means. The Canadian public inquiry model is also superior to that experienced in the United States where, as far as I understand it, events tend to channel controversies into the courts whenever possible where these controversies and their rational content dissipate amidst the richly rewarded sophistries of lawyers.

We have good evidence in the Canadian model that participation by citizens, by voluntary public interest groups, and by university researchers in such hearings has made a difference, however small, in the ultimate outcomes. This should spur us on to find ways of further improving the chances for informed participation.

References

Aidala, James V., Jr.
1979 "Regulating carcinogens: The case of pesticides." Paper presented at annual meetings of the Society for the Study of Social Problems, Boston, August.

Albrecht, Stan L., and A. L. Mauss
1975 "The environment as a social problem." In *Social Problems as Social Movements*, edited by A. L. Mauss, 556–605. Philadelphia: J. B. Lippincott.

Alexander, Tom
1981 "A simpler path to a cleaner environment." *Fortune* (4 May): 234ff.

Alford, Robert R., and R. Friedland
1975 "Political participation and public policy." *Annual Review of Sociology*: 429–79.

Anderson, Charles H.
1976 *The Sociology of Survival: Social Problems of Growth*. Homewood, Ill.: Dorsey Press.

Aristotle
1934 *The Nichomachean Ethics*. Translated by H. Rackham. London: W. Heinemann.

Arnold, Guy
1978 *Britain's Oil*. London: Hamish Hamilton.

Austin, J. L.
1962 *How to Do Things with Words*. Cambridge, Mass.: Harvard University Press.

Baker, Wayne, and R. Hertz
1980a "The dual opportunity structure in an Israeli kibbutz: Men's and women's work." Paper presented at Midwest Sociological Society meetings, Milwaukee, Wis.

1980b "The organization of work in a collectivist community." Paper presented at Midwest Sociological Society meetings, Minneapolis.

1981 "Communal diffusion of friendship: The structure of intimate relations in an Israeli kibbutz." In *Research in the Interweave of Social Roles*, edited by H. Z. Lopata and D. Maines. Vol. 2, Greenwich, Conn: JAI Press.

1982 "Women's and men's work in an Israeli kibbutz: The allocation of labor." In *Women in the Kibbutz*, edited by M. Rosner. Norwood, Pa: Norwood Press.

Bankes, Nigel, and Andrew R. Thompson
1981 "Monitoring for impact assessment and management: An analysis of the legal and administrative framework." Westwater Research Centre, Vancouver: University of British Columbia.
Barkan, Steven E.
1979 "Strategic, tactical and organizational dilemmas of the protest movement against nuclear power." *Social Problems* 27 (1): 19–37.
Barkley, Paul W., and D. W. Seckler
1972 *Economic Growth and Environmental Decay: The Solution Becomes the Problem*. New York: Harcourt Brace Jovanovich.
Barron, Ian, and Ray Curnow
1979 *The Future with Microelectronics*. New York: Nichols Publishing Co.
Bazelon, D.
1979 "Risk and responsibility." *Science* 205: 277–80.
Bell, Daniel
1973 *The Coming of Post-Industrial Society*. New York: Basic Books.
1976 "Foreword, 1976." In *The Coming of Post-Industrial Society*. New York: Basic Books.
1980 "The social function of the information society." In *The Microelectronics Revolution*, edited by T. Forester, 500–549. Oxford: Basil Blackwell.
Bjerring, A. K., and C. A. Hooker
1980 "The implications of philosophy of science for science policy." Paper prepared for Conference on the *Human Context for Science and Technology*. Ottawa: Social Sciences and Humanities Research Council of Canada.
Blau, Peter M., Cecillia McHugh Falbe, William McKinley, and Phelps K. Tracy
1976 "Technology and organization in manufacturing." *Administrative Science Quarterly* 21: 20–40.
Blauner, Robert
1964 *Alienation and Freedom: The Factory Worker and His Industry*. Chicago: University of Chicago Press.
Blumberg, Rae L.
1974 "From liberation to laundry: A structural interpretation of retreat from sexual equality in the Israeli kibbutz." Paper presented at American Political Science Association meetings, Chicago.
1977 "Women and work around the world." In *Beyond Sex Roles*, edited by A. G. Sargent, 412–33. St. Paul, Minn.: West.
Blumberg, Melvin, and Donald Gerwin
1981 "Human consequences relating to the acquisition and use of computerized manufacturing technology." Paper presented to the international conference on "QWL and the '80s," Toronto, 30 August–3 September.

Borow, Henry, ed.
1964 *Man in a World at Work.* Boston: Houghton Mifflin.
Bowes, Alison
1978 "Women in the kibbutz movement." *Sociological Review* 26: 237–62.
Bradley, Gunilla
1981 "Computerization: Psychosocial aspects." Paper presented to the international conference on "QWL and the '80s," Toronto, 30 August–3 September.
Braverman, Harry
1974 *Labor and Monopoly Capital: The Degradation of Work in the Twentieth Century.* New York: Monthly Review Press.
Brunet, Lucie
1980 "Word processors: Source of emancipation or alienation of office workers?" *Quality of Working Life* 3 (4): 17–19.
Burgoyne Report
1980 "Offshore safety." Cmnd. 8941 London: HMSO.
Buttel, Frederick H.
1976 "Social science and the environment: Competing theories." *Social Science Quarterly* 57 (2): 307–23.
——— and W. L. Flinn
1975 "Methodological issues in the sociology of natural resources." *Humboldt Journal of Social Relations* 3 (1): 63–72.
Butteriss, Margaret
1981 "The organizational and social factors involved in introducing office automation." Paper presented to the international conference on "QWL and the '80s," Toronto, 30 August–3 September.
Campbell, H., and A. Scott
1978 "Postponement v. Attenuation. An Analysis of Two Strategies for Predicting and Mitigating the Environmental Damage of Large-Scale Uranium Mining Projects." Vancouver, University of British Columbia: Programme in Natural Resource, Westwater Institute.
Canadian Broadcasting Corporation
1982 "Morningside Radio Show." Host Don Harron, Interview on the *Ocean Ranger* Disaster. 19 February.
Canadian Press
1981 "Women workers face tough times in next decade." *The Calgary Herald* (January).
Carson, Rachel
1962 *Silent Spring.* New York: Houghton Mifflin.
Carson, W. G.
1981 *The Other Price of Britain's Oil.* Oxford: Martin Robertson.
Chamberlain, Neil W.
1977 *Remaking American Values.* New York: Basic Books.

Chandler, Alfred, Jr., and Herman Daems, eds. and contributors
1981 *Managerial Hierarchies: Comparative Perspectives on the Rise of Modern Industrial Enterprise*. Cambridge, Mass.: Harvard University Press.
Child, John
1973 "Predicting and understanding organizational structure." *Administrative Science Quarterly* 18: 168–85.
Cicero
1913 *De Officiis*. Book I, xlii. Loeb Classical Library. London: Heinemann; New York: Macmillan.
Cicourel, A. V.
1973 *Cognitive Sociology: Language and Meaning in Social Interaction*. Middlesex: Penguin.
Clark, S., ed.
1981 *Environmental Assessment in Australia and Canada*. Vancouver: Westwater Research Centre.
Committee of Inquiry into Technological Change (Australia)
1980 Report ["Myers Report"]. Canberra: Australian Government Printing Service.
Committee of Public Accounts, U.K.
1973 "North Sea oil and gas." *Parliamentary Papers*, 1972–73, Volume 14. London: HMSO.
Committee to Review the Functioning of Financial Institutions
1978 *The Financing of North Sea Oil*. London: HMSO.
Commoner, Barry
1970 *Science and Survival*. New York: Ballantine Books.
1977 *The Poverty of Power: Energy and the Economic Crisis*. New York: Bantam.
Commons, John R.
1961 *Institutional Economics*, Volumes I and II. Madison, Wis.: University of Wisconsin Press.
Congressional Quarterly
1981 *Energy Policy*. 2d ed. Washington: D.C.: Congressional Quarterly, Inc.
Cooper, Charles
1973 "Choice of technique and technological change as problems in political economy." *International Social Science Journal* 25 (3): 293–304.
Council on Environmental Quality
1980 *Public Opinion on Environmental Issues: Results of a National Public Opinion Survey*. Washington, D.C.: U.S. Government Printing Office.
Crain, R., E. Katz, and D. Rosenthal
1969 *The Politics of Community Conflict*. Indianapolis: Bobbs-Merrill.
Crandall, Robert W., and Lester B. Lave, eds.
1981 *The Scientific Basis of Health and Safety Regulation: Studies in the Regulation of Economic Activity*. Washington, D.C.: Brookings Institution.

D'Aquino, Thomas, et al.
1979 *Parliamentary Government in Canada: A Critical Assessment and Suggestions for Change*. Ottawa: Intercounsel Ltd.

Darroch, Robert
1980 "The brave new world built on silicon chips." *The Bulletin Australia* (8 January): 37–43.

del Sesto, Steven
1980 "Science, sensationalism and the media." Unpublished. New York: Russell Sage Foundation.

de Moll, Lane
1978 "No more 'rich tech, poor tech.' " *Ozarka* (special issue): 3.

Department of Energy, UK
1974 *Development in the Oil and Gas Resources of the United Kingdom*. London: HMSO.

Dewees, Donald N.
1981 *Evaluation of Policies for Regulating Environmental Pollution*. Working Paper No. 4. Ottawa: Economic Council of Canada.

Dickson, David
1974 *Alternative Technology and the Politics of Technical Change*. Glasgow: Fontans/Collins.
1981 "Limiting democracy: Technocrats and the liberal state." *Democracy* 1 (1): 61–79.

Doern, Bruce G.
1981 *The Peripheral Nature of Scientific and Technical Controversy in Federal Policy Formation*. Ottawa: Science Council of Canada Background Study 46.

Dorcey, A. H. J.
1983 "Coastal zone management as a bargaining process." *Coastal Zone Management Journal* 11 (1–2): 13–40.

———, Michael W. McPhee, and Sam Sydneysmith
1981 *Salmon Protection and the B.C. Coastal Forest Industry: Environmental Regulation as a Bargaining Process*. Vancouver: Westwater Research Centre, University of British Columbia.

Dunette, Marvin D.
1973 *Work and Nonwork in the Year 2001*. Monterey, Calif.: Brooks/Cole Publishing Co.

Economic Council of Canada
1979 *Fifteenth Annual Review: A Time for Reason*. Ottawa.

Eddy, Howard R.
1981 *Sanctions, Compliance Policy and Administrative Law*. Draft study, Ottawa: Law Reform Commission of Canada.

Ehrlich, Paul
1968 *The Population Bomb*. New York: Ballantine.

Ellul, Jacques
1967 *The Technological Society*. New York: Alfred A. Knopf. (1st ed., 1964).

Elson, Peter
1982 "Office automation: A weighty problem." *The Financial Post* (Canada) (3 April): 14.

Etzioni, Amitai
1964 *Modern Organizations*. Englewood Cliffs, N.J.: Prentice-Hall.

Fairfax, Sally K.
1978 "A disaster in the environmental movement." *Science* 199 (17 February): 743–48.

Faunce, William
1968 *Problems of an Industrial Society*. Englewood Cliffs, N.J.: Prentice-Hall.
1970 "Automation and the division of labor." In *Automation, Alienation, and Anomie*, edited by S. Marcson. New York: Harper and Row.

Felske, Brian E.
1981 *Sulphur Dioxide Regulation and the Canadian Nonferrous Metals Industry*. Technical Report No. 3, Ottawa: Economic Council of Canada.

Feree, M. M., and F. D. Miller
1980 "Mobilization and meaning: Toward an integration of social psychological and resource perspectives on social movements." Unpublished manuscript.

Ferguson, John
1982 "Union seeks fewer hours." *The Calgary Herald* (22 April): A20.

Feyerabend, Paul K.
1975 *Outline of an Anarchistic Theory of Knowledge*. Highlands, N.Y.: Humanities Press.

Fireman, B., and W. A. Gamson
1977 "Utilitarian logic in the resource mobilization perspective." Unpublished manuscript.

Fisher, Harold
1954 "Big things ahead." *American Magazine* 157 (April): 21.

Flood, Michael, and Robin Grove-White
1976 *Nuclear Prospects: A Comment on the Individual, the State and Nuclear Power*. London: Friends of the Earth.

Frahm, Annette, and F. H. Buttel
1980 "Appropriate technology: Current debate and future possibilities." Unpublished, Department of Rural Sociology, Cornell University.

Frank, Charles
1982 "Who wants the electronic office?" *The Calgary Herald* (5 April): D2.

Freeman, Christopher
1978 "Technical change and future employment prospects in industrialized societies." Paper presented at a Six Countries Workshop on Government Policies Towards Technological Innovation in Industry, Paris, France, 13–14 November.

Friedland, Roger, F. F. Piven, and R. R. Alford
1977 "Political conflict, urban structure, and the fiscal crisis." *International Journal of Urban and Regional Research* 1 (3): 447–71.

Friedman, Judith J.
1980 "Environmental policy, equity, and sociology." Paper presented at annual meetings of the American Sociological Association, New York, August.

Friesema, H. Paul, and P. J. Culhane
1976 "Social impacts, politics, and the environmental impact statement process." *Natural Resources Journal* 16 (April): 339–56.

Fromm, E.
1961 *Marx's Concept of Man*. New York: Unger.

Fuller, John
1975 *We Almost Lost Detroit*. New York: Ballantine.

Galbraith, John Kenneth
1967 *The New Industrial State*. Boston: Houghton Mifflin.
1978 *The New Industrial State*. 3d revised ed. New York: Mentor.

Gallie, Duncan
1978 *In Search of the New Working Class: Automation and Social Integration Within the Capitalist Enterprise*. Cambridge: Cambridge University Press.

Gamson, William A.
1975 *The Strategy of Social Protest*. Homewood, Ill.: Dorsey Press.
1980 "Understanding the careers of challenging groups: A commentary on Goldstone." *American Journal of Sociology* 85 (5): 1043–60.

Gendron, Bernard
1977 *Technology and the Human Condition*. New York: St. Martin's Press.

Geriffi, Gary
1978 "Drug firms and dependency in Mexico: The case of the steroid drug industry." *International Organizations* 32: 237–86.

Gershuny, Jonathan
1978 *After Industrial Society? The Emerging Self-Service Economy*. London: Macmillan.

Goldstone, Jack A.
1980 "The weakness of organization: A new look at Gamson's *The Strategy of Social Protest*." *American Journal of Sociology* 85 (5): 1017–42.

Government of Canada
1979 *Royal Commission on Financial Management and Accountability Final Report* (Lambert Commission). Ottawa: Ministry of Supply and Services.

Gray, Alexander
1951 *The Development of Economic Doctrine*. London: Longmans, Green and Co.

Habermas, J.
1976 *Legitimation Crisis*. London: Heinemann.

Hall, Oswald, and Richard Carlton
1977 *Basic Skills at School and Work*. Toronto: Ontario Economic Council.

Hamilton, A.
1978 *North Sea Impact*. London: International Institute for Economic Research.

Hannon, Bruce
1975 "Energy conservation and the consumer." *Science* 189 (11 July): 95–102.

Hardin, Garrett
1968 "The tragedy of the Commons." *Science* 162: 1243.

Harris, L.
1981 "Inflation seen as most important problem facing country." *The Harris Survey* (25 May): 42.

Harvey, Edward B.
1975 *Industrial Society: Structure, Roles and Relations*. Homewood, Ill.: Dorsey Press.

Hawkins, John D.
1975 "Utopian values and communal social life: A comparative study of social arrangements in four counter culture communes established to realize participants' values." Unpublished doctoral thesis, Department of Sociology. Northwestern University, Evanston, Illinois.

Hays, Samuel P.
1969 *Conservation and the Gospel of Efficiency: The Progressive Conservation Movement, 1890–1920*. New York: Atheneum.

Henderson, Hazel
1978 *Creating Alternative Futures: The End of Economics*. New York: Berkeley Publications Corp.

Hickson, David J., D. S. Pugh, and Diane Pheysey
1969 "Operations technology and organizational structure: An empirical reappraisal." *Administrative Science Quarterly* 14: 378–97.

Hirsch, J.
1978 "The state apparatus and social reproduction." In *State and Capital*, edited by T. Holloway and S. Piccioto. London: Edward Arnold.

Hirsch, Paul M.
1969 *The Structure of the Popular Music Industry*. Ann Arbor, Mich.: University of Michigan Survey Research Center.

Hooker, C. A., ed.
1980 *The Human Context for Science and Technology*, Final Report, Vol. 1. Ottawa: Social Sciences and Humanities Research Council of Canada.

Hooker, Clifford A., et al.
1981 *Energy and the Quality of Life: Understanding Energy Policy*. Toronto: University of Toronto Press.

Hornback, Kenneth E.
1974 "Orbits of opinion: The role of age in the environmental movement's attentive pubic, 1968–1972." Unpublished doctoral dissertation, Department of Sociology, Michigan State University, East Lansing, Michigan.

House, J. Douglas
1980 *The Last of the Free Enterprisers: The Oilmen of Calgary*. Toronto: Macmillan.

Humphrey, Craig R., and F. R. Buttel
1982 *Environment, Energy, and Society*. Belmont, Calif.: Wadsworth.

Hunt, Constance D., and Alastair R. Lucas
1981 *Environmental Regulation—Its Impact on Major Oil and Gas Projects: Oil Sands and Arctic*. Calgary: Canadian Institute of Resources Law, University of Calgary.

Hutchinson, Gordon
1982 "Canada's robot sector still in embryonic stage." *Toronto Globe and Mail* (29 March).

Inglehart, Ronald
1977 *The Silent Revolution: Changing Values and Political Styles Among Western Publics*. Princeton, N.J.: Princeton University Press.

Kitchen, J.
1977 *Labour Law and Offshore Oil*. London: Croom Helm.

Knickerbocker, Frederick K.
1973 *Oligopolistic Reaction and Multinational Enterprise*. Boston: Division of Research, Graduate School of Business Administration, Harvard University.

Kolankiewicz, Leon
1981 "Implementation of British Columbia Pollution Control Act, 1967, in the Lower Fraser River." Vancouver: University of British Columbia, Unpublished M.Sc. thesis.

Kuhn, Thomas S.
1970 *The Structure of Scientific Revolutions*. 2d ed. Chicago: University of Chicago Press.

Kumar, Krishan
1978 *Prophecy and Progress*. Harmondsworth: Penguin Books.

Kuznets, Simon
1953 *Economic Change*. New York: W. W. Norton.

Lamberton, D. McL.
1980 "Computers and jobs." Presented to the Industrial Relations Section of the 50th ANZAAS Congress, Adelaide (May).

Lane, Robert E.
1966 "The decline of politics and ideology in a knowledgeable society." *American Sociological Review* 31: 649–62.
Lansbury, Russell D.
1980 "New technology and industrial relations in the retail grocery industry: Some lessons for Australia from international experience." Presented to the Industrial Relations Section of the 50th ANZAAS Congress, Adelaide (May).
Largey, G. P., and D. R. Watson
1972 "The sociology of odors." *American Journal of Sociology* 77, no. 6 (May): 1021–34.
Larkin, Peter A.
1979 *Fisheries Management: The Coming Crisis, Coastal Resources in the Future of B.C.*. Vancouver: Westwater Research Centre, University of British Columbia.
Lasswell, Harold D.
1948 "The structure and function of communication in society." In *The Communication of Ideas*, edited by L. Bryson. New York: Harper and Brothers.
Laurence, William
1966 "Is atomic energy the key to our dreams?" *Saturday Evening Post* 222 (March–April): 41.
Lave, Lester B., and E. P. Seskin
1970 "Air pollution and human health." *Science* 169 (21 August): 723–33.
Law Reform Commission of Canada
1977 *Our Criminal Law*. Ottawa: Supply and Services.
Levison, Andrew
1974 "The working-class majority." *New Yorker* (2 September): 36–61.
Lifton, Robert
1980 *The Broken Connection*. New York: Alfred A. Knopf.
Lindblom, Charles E.
1977 *Politics and Markets: The World's Political-Economic Systems*. New York: Basic Books.
Lovins, Amory
1976 "Energy strategy: The road not taken?" *Foreign Affairs* 55(1): 65–96.
1977 *Soft Energy Paths: Toward a Durable Peace*. New York: Harper Colophon.
Lowi, Theodore J.
1964 "American business, public policy, case-studies, and political theory." *World Politics* 16 (4): 677–715.
1972 "Four systems of policy, politics, and choice." *Public Administration Review* 32 (4): 298–310.
1979 *The End of Liberalism*. 2d ed. New York: W. W. Norton.

Marchak, P.
1979 *In Whose Interest?* Toronto: McClelland and Stewart.
Marcuse, Herbert
1964 *One Dimensional Man*. Boston: Beacon Press.
Marx, Karl
1969 *Theories of Surplus Value* (Volume 4 of *Capital*), Part I. Burns translation. Moscow: Progress Publishers.
Marx, Leo
1980 "American literary culture and the fatalistic view of technology." *Alternative Futures* 3 (2): 48–49.
Mazur, A.
1977 "Science Courts." *Minerva* 15 (April): 1–14.
1981a *The Dynamics of Technical Controversy*. Washington, D.C.: Communications Press.
1981b "Unsuspected bias in the evenhanded reporting of technical controversies." Syracuse University Sociology Department, unpublished.
———, and B. Conant
1978 "Opposition to a local nuclear waste repository." *Social Studies of Science* 8: 235–43.
McCarthy, J. D., and M. N. Zald
1977 "Resource mobilization and social movements: A partial theory." *American Journal of Sociology* 82: 1212–41.
McClelland, D. C.
1961 *The Achieving Society*. New York: Van Nostrand.
McNeil, Kenneth
1978 "Understanding organizational power: Building on the Weberian Legacy." *Administrative Science Quarterly* 25: 65–90.
McNulty, J.
1981 "The economic basis for an information society." Department of Communication, Simon Fraser University, unpublished.
McPhee, Michael W.
1978 "Water quality management in British Columbia: Pollution control in the pulp and paper industry." Burnaby, B.C.: Simon Fraser University, unpublished M.A. thesis.
Mead, Margaret
1967 *The Changing Cultural Patterns of Work and Leisure*. Washington, D. C.: U.S. Department of Labor.
Meidinger, Errol, and A. Schnaiberg
1980 "Social impact assessment as evaluation research: Claimants and claims." *Evaluation Review* 4 (4): 507–35.
Menville, Douglas
1975 *A History and Critical Survey of the Science Fiction Film*. New York: Arno Press.

Menzies, Heather
1981 *Women and the Chip: Case Studies of the Effects of Informatics on Employment in Canada*. Montreal: Institute for Research on Public Policy.

Merton, Robert K.
1957 *Social Theory and Social Structure*. Revised ed. New York: Free Press.

Mesthene, E. G.
1967 "Technology and wisdom." In *Technology and Social Change*. Indianapolis: Bobbs-Merrill.
1970 *Technological Change: Its Impact on Man and Society*. Cambridge, Mass.: Harvard University Press.

Mills, C. Wright
1959 *The Sociological Imagination*. Oxford: Oxford University Press.

Mishan, Ezra J.
1967 *The Costs of Economic Growth*. New York: Frederick A. Praeger.
1973 "Ills, bads, and disamenities: The wages of growth." In *The No-Growth Society*, edited by M. Olson and H. H. Landsberg, 63–87. New York: W. W. Norton.

Mitchell, Robert C.
1978 "Environment: An enduring concern." *Resources* 57 (January–March): 1ff.
1979 "Public opinion and nuclear power before and after Three Mile Island." *Resources* (January–April): 5–7.

Molotch, Harvey, and M. Lester
1974 "News as purposive behavior: On the strategic use of routine events, accidents, and scandals." *American Sociological Review* 39 (1): 101–12.

Morris, David, and K. Hess
1975 *Neighborhood Power: Returning Political and Economic Power to Community Life*. Boston: Beacon Press.

Morrison, Denton E.
1980 "The soft, cutting edge of environmentalism: Why and how the A-T notion is changing the movement." *Natural Resources Journal* 20 (2): 275–98.

———, and D. G. Lodwick
1981 "The social impacts of soft and hard energy systems: The Lovins's claims as a social science challenge." *Annual Review of Energy* 6: 357–78.

Morse, Dean, and Susan H. Gray
1980 *Early Retirement: Boon or Bane? A Study of Three Large Corporations*. Montclair, N.J.: Allanheld, Osmun.

Mueller, J.
1968 "Fluoridation attitude change." *American Journal of Public Health* 58: 1876.

Mumford, Lewis
1973 *Technics and Civilization*. New York: Harcourt, Brace, and World.
Myrdal, Gunnar
1968 *Asian Drama: An Inquiry into the Poverty of Nations*. New York: Pantheon.
Nelkin, Dorothy, and S. Fallows
1978 "The evolution of the nuclear debate: The role of public participation." *Annual Review of Energy* 3: 275–312.
Nelson, J. G., J. C. Day, and Sabine Jessen
1981 *Environmental Regulation of the Nanticoke Industrial Complex*. Working Paper No. 7. Ottawa: Economic Council of Canada.
Nelson, Ruben
1980 *The Illusions of Urban Man*. 2d ed. Ottawa: Square One Management Ltd.
Nemetz, Peter, John Sturdy, Dean Uyeno, Patricia Vertinsky, Ilan Vertinsky, and Aidan Vining
1981 *Regulation of Toxic Chemicals in the Environment*. Working Paper No. 20. Ottawa: Economic Council of Canada.
No on 15 Committee
1976 *California Energy Bulletin*. Glendale, Calif.: No on 15 Committee/Californians Against the Nuclear Shutdown.
Noble, David F.
1979 *America by Design: Science, Technology, and the Rise of Corporate Capitalism*. New York: Alfred A. Knopf.
Nora, Simon, and Alain Minc
1979 *L'informatisation de la Societe*. Paris: La Documentation Francaise.
1980 *The Computerization of Society*. Cambridge, Mass.: MIT Press.
Nordhaus, W., and J. Tobin
1972 "Is growth obsolete?' In *Economic Growth*. New York: BNER.
Nore, Petter
1976 *Six Myths of British Oil Politics*. London: Thames Papers in Political Economy.
1978 "International oil in Norway." In *The Nationalisation of Multinationals in Peripheral Economies*, edited by J. Faundez and S. Piccioto, 168. London: Macmillan.
Noreng, O.
1980 *The Oil Industry and Government Strategy in the North Sea*. London: Croom Helm.
Nosow, Sigmund, and William H. Form, eds.
1962 *Man, Work and Society*. New York: Harper.
O'Connor, James
1973 *The Fiscal Crisis of the State*. New York: St. Martin's.
Offshore Services
1973. July (UK)

Ogburn, William F.
1922 *Social Change*. New York: Heubsch.

Okun, Arthur
1975 *Equality and Efficiency: The Big Tradeoff*. Washington, D.C.: Brookings Institution.

Olsen, Mancur
(1965) *The Logic of Collective Action*. Cambridge, Mass.: Harvard University Press.

Owen, John D.
1969 *The Price of Leisure*. Montreal: McGill-Queen's University Press.

Ozark Institute
1978 "A special report on A-T." *Ozarka*, Journal of the Ozark Institute (special issue).

Paden-Eisenstark, Dorit
1973 "Are Israeli women really equal? Trends and patterns of Israeli women's labor force participation—a comparative analysis." *Journal of Marriage and the Family* 35: 538–45.

Parish, James, and Michael Pitts
1977 *The Great Science Fiction Pictures*. Metuchen, N.J.: Scarecrow Press.

Parliamentary Task Force on Employment Opportunities for the 80's
1981 *Work for Tomorrow*. Ottawa: Speaker of the House of Commons.

Parrish, William L., and M. K. Whyte
1978 *Village and Family in Contemporary China*. Chicago: University of Chicago Press.

Penner, Peter
1978 "Taking A-T to task." *Ozarka* (special issue): 2.

Perkins, William
1977 "Comments" to Atomic Industrial Forum before the International Science Fiction Convention, Miami, Florida, 4 September, 1977.

Pitts, James P.
1981 "Urban neighborhoods and A-T: A case study in technology transfer." Paper presented at annual meetings of the American Sociological Association, Toronto, Canada, August.

Porat, M. U.
1977 *The Information Economy*. Washington, D.C.: Office of Telecommunications, Special Publication 77–12. 9 volumes.

Purdy, David
1976 "British capitalism since the war." *Marxism Today* 20 (9 and 10): 270–77, 310–18.

Rabin, A. I.
1970 "The sexes: Ideology and reality in the Israeli kibbutz." *Sex Roles in Changing Society*, edited by G. H. Seward and R. C. Williamson, 285–307. New York: Random House.

Rankin, Murray, and Peter Finkle
1981 "The enforcement of environmental law: Taking the environment seriously." Unpublished paper. University of Victoria.
Rayman, Paula
1977 "Community development and nation-building: The study of an Israeli border kibbutz." Boston, Mass.: Boston College, Unpublished doctoral thesis.
Regina Group for a Non-Nuclear Society
1980 *Why People Say No*. Regina: Regina Group for a Non-Nuclear Society.
Reich, Charles A.
1970 *The Greening of America*. New York: Random House.
Rice, Berkeley
1982 "The Hawthorne defect: Persistence of a flawed theory." *Psychology Today* (February): 71–74.
Rinehart, James W.
1975 *The Tyranny of Work*. Don Mills, Ontario: Longman Canada Ltd.
Ritzer, George
1977 *Working: Conflict and Change*. 2d ed. Englewood Ciffs, N.J.: Prentice-Hall.
Robertson, James
1978 *The Sane Alternative: Signposts to a Self-Fulfilling Future*. London: James Robertson.
Rodgers, Daniel T.
1978 *The Work Ethic in Industrial America, 1850–1920*. Chicago: University of Chicago Press.
Rohlich, Gerard A., and Richard Howe
1981 *The Toxic Substances Control Act; Overview and Evaluation*. Working Paper No. 21. Ottawa: Economic Council of Canada.
Rosencranz, Armin
1981 "Economic approaches to air pollution control." *Environment* 23 (8): 25–30.
Rossin, A. David
1980 "The soft energy path: Where does it really lead?" *The Futurist* 14 (3): 57–63.
Rothschild-Whitt, Joyce
1979 "The collectivist organization: An alternative to rational-bureaucratic models." *American Sociological Review* 44 (August): 509–27.
Rubin, D., et al.
1979 *The Accident at Three Mile Island: Report of the Public's Right to Information Task Force*. Washington, D.C.: U.S. Superintendent of Documents.
Salter, Liora, and Debra Slaco
1981 *Public Inquiries in Canada*. Ottawa: Science Council of Canada.

Sandman, P., and M. Paden
1979 "At Three Mile Island." *Columbia Journalism Review* (July/August): 43–58.
Sapolsky, H.
1968 "Science, voters and the fluoridation controversy." *Science* 162: 427.
Schlesinger, Yaffa
1977 "Sex roles and social change in the kibbutz." *Journal of Marriage and the Family* 31: 567–83.
Schnaiberg, Allan
1973 "Politics, participation and pollution: The 'environmental' movement." In *Cities in Changes: Studies on the Urban Condition*, edited by J. Walton and D. E. Carns, 605–27. Boston: Allyn and Bacon.
1975 "Social syntheses of the societal-environmental dialectic: The role of distributional impacts." *Social Science Quarterly* 56 (June): 5–20.
1980 *The Environment: From Surplus to Scarcity*. New York: Oxford University Press.
1981 "Energy, equity and austerity: Some political impacts of energy rationing by price." Paper presented at annual meeting of Society for the Study of Social Problems, Toronto, Canada, August.
1982a "Energy conservation in the U.S.: A pyrrhic social victory?" Draft manuscript, Department of Sociology, Northwestern University, January.
1982b "Soft energy and hard labor? Structural restraints on the transition to A-T." Forthcoming in *Technology and Social Change*, edited by G. Summers et al.
1982c "Saving the environment: From whom, for whom, and by whom?" Forthcoming in *Environmental Politics and Policies: An International Perspective*, edited by N. Watts and P. Knoepfel.
Schon, Donald
1967 *Technology and Change: The New Heraclitus*. New York: Delacorte Press.
Schultz, Richard
1980 *Federalism, Bureaucracy, and Public Policy*. Montreal: McGill-Queens University Press.
Schumacher, E. F.
1973 *Small Is Beautiful: Economics as If People Mattered*. New York: Harper and Row.
1977 *A Guide for the Perplexed*. New York: Harper and Row.
1979 *Good Work*. New York: Harper Colophon.
Schumpeter, Joseph A.
1947 *Capitalism, Socialism, and Democracy*. 3d ed. New York: Harper and Brothers.

Science Council of Canada

1981 *Certainty Unmasked: Science, Values and Regulatory Decision Making.* Ottawa: A draft report of the Science Council of Canada Committee on Science and the Legal Process.

1983 *Planning Now for an Information Society.* Ottawa: Minister of Supply and Services, 1982.

Science Council of Canada/IRPP Workshop

1980 *Biotechnology in Canada: Promises and Concerns.* Ottawa: Science Council of Canada.

Scitovsky, Tibor

1977 *The Joyless Economy.* Oxford: Oxford University Press.

Scott, J.

1979 *Corporations, Classes and Capitalism.* London: Croom Helm.

Scott, Robert A., and Arnold R. Shore

1979 *Sociology Does Not Apply: A Study of the Use of Sociology in Public Policy.* New York: Elsevier.

Sea Gem Report

1967 *Report of the Tribunal Appointed to Inquire into the Causes of the Accident to the Drilling Rig Sea Gem.* Cmnd. 3409. London: HMSO.

Selznick, Philip

1966 *TVA and the Grass Roots: A Study in the Sociology of Formal Organization.* New York: Harper.

Senior, N. W.

1837 *Letters on the Factory Act.* London.

Serafini, Shirley, and Michel Andrieu

1981 *The Information Revolution and Its Implications for Canada.* Hull, Quebec: Communications Economics Branch, Department of Communications, Canadian Government.

Shaheen, Jack, ed.

1978 *Nuclear War Films.* Carbondale, Ill.: Southern Illinois Press.

Shepard, Jon

1971 *Automation and Alienation: A Study of Office and Factory Workers.* Cambridge, Mass.: MIT Press.

Shils, Edward

1970 "Automation: Technology or concept." In *Automation, Alienation and Anomie,* edited by Simon Marcson. New York: Harper and Row.

Sigelman, Lee

1973 "Reporting the news: An organizational analysis." *American Journal of Sociology* 79 (1): 132–51.

Smith, Eleanor

1977 "Keeping the lid on anti-nuclear films." *In These Times* (6–12 September).

Smith, P. L., et al.
1969 *The Measurement of Satisfaction in Work and Retirement*. Chicago: Rand McNally.

Soddy, Frederick
1909 "The enemy of radium." *Hughes Monthly* (20 December): 52–59.

Spector, Malcolm, and J. I. Kitsuse
1977 *Constructing Social Problems*. Menlo Park, Calif.: Cummings Publishing Co.

Sproule-Jones, Mark
1981 *The Real World of Pollution Control*. Vancouver: Research Centre, University of British Columbia.

Stanfield, Hon. Robert
1978 "The present state of the legislative process in Canada: Myths and realities." In *The Legislative Process in Canada: The Need for Reform*, edited by William A. Neilson and James C. MacPherson. Montreal: Institute for Research on Public Policy.

Statutory Instruments, UK
1974–78 London: HMSO.

Steinhart, Jim
1982 "Word processors sales booming during economic slump." *The Globe and Mail* (29 March): R3.

Stinchcombe, Arthur L.
1965 "Social structure and the founding of organizations." In *Handbook of Organizations*, edited by J.G. March, 153–69. Chicago: Rand McNally.

Stopford, John M., and Louis T. Wells, Jr.
1972 *Managing the Multinational Enterprise: Organization of the Firm and Ownership of Subsidiaries*. New York: Basic.

Stretton, Hugh
1976 *Capitalism, Socialism and the Environment*. Cambridge: Cambridge University Press.

Swaigen, John Z.
1981 *Compensation of Pollution Victims in Canada*. Ottawa: Economic Council of Canada.

Talbot, David, and R. E. Morgan
1981 "Soft energy for hard times." *Environmental Action* (October): 26–29.

Talmon, Yonina
1972 *Family and Community in the Kibbutz*. Cambridge, Mass.: Harvard University Press.

Task Force of the Presidential Advisory Group on Anticipated Advances in Science
1976 "The science court experiment: An interim report." *Science* 193: 653–56.

Taylor, James C.
1981 "Employee participation in the socio-technical systems analysis of a computer operations organization." Paper presented at "QWL and the '80's" Conference, Toronto, Canada (30 August–3 September, 1981).
Taylor, Overton H.
1960 *A History of Economic Thought*. Toronto: McGraw-Hill.
Thompson, Andrew R.
1981 *Environmental Regulation in Canada: An Assessment of the Regulatory Process*. Vancouver: Westwater Research Centre, University of British Columbia.
Thompson, G. B.
1979 "Memo from Mercury: Information technology is different." Institute for Research on Public Policy, Occasional Paper No. 10.
Thurow, Lester C.
1980 *The Zero-Sum Society: Distribution and the Possibilities for Economic Change*. New York: Basic Books.
Tiger, Lionel, and J. Shepher
1975 *Women in the Kibbutz*. New York: Harcourt Brace Jovanovich.
Toffler, Alvin
1981 *The Third Wave*. New York: Bantam Books.
Tuchman, G.
1978 *Making News*. New York: Free Press.
Tucker, William
1981 "Marketing pollution." *Harpers* (May): 31–38.
Ure, A.
1835 *The Philosophy of Manufactures*. London.
Utredninger, Norges Offentlige
1981 *"Alexander L. Kielland"—ulykken, Nov. 1981: 11*. Oslo: Universitetsforleget.
Valaskakis, Kimon
1979 *Information Society Project*. Montreal: GAMMA.
Vernon, Raymond
1966 "International investment and international trade in the product cycle." *Quarterly Journal of Economics* 80: 190–207.
Vickers, Sir Geoffrey
1970 *Freedom in a Rocking Boat: Changing Values in an Unstable Society*. Baltimore, Md.: Penguin Books.
Victor, P., and T. Burrell
1981 "Environmental protection regulation, water pollution, and the pulp and paper industry." Ottawa: Economic Council of Canada.
Walsh, Edward J.
1981 "Resource mobilization and citizen protest in communities around Three Mile Island." *Social Problems* 29 (1): 1–21.

Weart, Spencer R.
1979 *Scientists in Power*. Cambridge, Mass.: Harvard University Press.
1980 "Nuclear fear: A history and an experiment." Unpublished manuscript. New York: American Institute of Physics.

Wells, H. G.
1914 *The World Set Free: A Story of Mankind*. North Hollywood, Calif.: Leisure Books Reprint.

White, I., et al.
1979 *North Sea Oil and Gas*. Norman: University of Oklahoma Press.

White, Terrence
1979a *Human Resource Management—Changing Times in Alberta*. Edmonton: Alberta Labour.
1979b "A survey of innovative work arrangements in Alberta." *Quality of Working Life* 2 (4): 20–22.

Williamson, Oliver E.
1981 "Emergence of the visible hand: Implications for industrial organization." In *Managerial Hierarchies*, edited by Alfred L. Chandler and Herman Daems, 182–202. Cambridge, Mass.: Harvard University Press, 1966.

Wolfbein, Seymour
1970 "The pace of technological change and the factors affecting it." In *Automation, Alienation and Anomie*, edited by Simon Marcson. New York: Harper and Row.

Wolin, Sheldon S.
1981 "The people's two bodies." *Democracy* 1 (1): 9–24.

Woodward, Joan
1965 *Industrial Organization: Theory and Practice*. Oxford: Oxford University Press.

Young, Colin
1962 "Nobody dies/patriotism in Hollywood." In *Film: Book 2/Films of Peace and War*, edited by Robert Hughes. New York: Grove Press.

Index

About the Editors and Contributors

AUGUSTINE BRANNIGAN is Associate Professor in the Department of Sociology at The University of Calgary, where he teaches in the areas of criminal justice and sociological theory. He is author of *The Social Basis of Scientific Discoveries* and has written an overview of the administration of justice in Canada.

HARVEY BUCKMASTER is Professor of Physics at The University of Calgary. He is a member of the NSERC Mainline Physics Grant Selection Committee, the NRC Pacific Science Association Committee, the Science Advisory Committee to the Environment Council of Alberta, and was awarded the Queen Elizabeth II Silver Anniversary Medal for "service to Canada" in 1978. He has written on the impact of science and technology on the environment and society, as well as having published over one hundred papers on the theoretical, experimental, and instrumental aspects of electron paramagnetic resonance and microwave dielectric spectroscopy.

W.G. CARSON is Professor and Chairperson of the Department of Legal Studies at La Trobe University in Melbourne, Australia. He has been a member of the Law Faculty at the University of Edinburgh, Scotland and the Faculty of Social Sciences at the University of London. Author of *The Other Price of Britain's Oil,* he has also written extensively in the field of factory legislation in the United Kingdom and elsewhere.

JOSEPH DiSANTO is Associate Professor of Sociology at The University of Calgary. His current research and publications are in the area of sociology of resource development and environmental impact assessment. He is a principal in HERA Consulting Ltd., which concentrates on social impact assessments for major resource developments.

PHILIP ELDER is Professor and Associate Dean of the Faculty of Environmental Design at The University of Calgary. His research interests and articles include aspects of environmental and planning law and policy. He has edited *Environmental Management and Public Participation* and is the author of *Soft is Hard: Barriers and Incentives in Canadian Energy Policy*.

SHELDON GOLDENBERG is Associate Professor in the Department of Sociology at The University of Calgary. He specializes in social impact assessment, methodology, social networks, and the sociology of knowledge. He has published many articles and is currently at work on two books.

WILLIAM LEISS is Professor and Chairman, Department of Communication, at Simon Fraser University. He has worked as a consultant with departments of the Federal Government in Canada on environmental and regulatory issues. He is the author of *Limits to Satisfaction* and *The Domination of Nature*; editor of *Ecology versus Politics in Canada*; and has contributed articles in the areas of technology and society, the theory of human needs, and social theory. He is currently acting as co-author to the forthcoming work, *Social Communication in Advertising*.

ALLAN MAZUR is Professor in the Faculty of Citizenship and Public Affairs at Syracuse University. He is a technologist and sociologist and worked for several years as an aerospace engineer. He has been a member of the social science faculties of MIT and Stanford University, and is a frequent lecturer and consultant to government agencies and private groups. He is the author of *The Dynamics of Technical Controversy*, in addition to many other publications.

DONALD L. MILLS is Professor of Sociology and Associate Dean of General Studies at The University of Calgary. He was a Research

Director for the Royal Commission on Health Services, and is author of *Chiropractic, Naturopathy and Osteopathy in Canada.* He is co-editor of *Professionalization*, and author of several articles on worklife. His current research interests include both dual-career couples and the Canadian legal profession.

DOROTHY NELKIN is Professor in the Faculty of the Program on Science, Technology and Society and the Department of Sociology at Cornell University. She is on the Board of Directors of the AAAS and the Council for the Advancement of Science Writing. Her research focuses on controversial areas of science and technology. She is the author of *The Atom Besieged, Workers at Risk, Controversy: The Politics of Technical Decisions*, and *The Creation Controversy*.

ALLAN OLMSTED is Associate Professor of Sociology at The University of Calgary. His research interests include mass communications, popular culture, human ecology, and the social impact of industrial development.

STEPHEN G. PEITCHINIS is Professor and Head of the Department of Economics at The University of Calgary. He has served as Chairman of a commission that investigated the financing of post-secondary education in Canada; Associate Director of the Human Resources Research Council of Alberta; and Dean of the Faculty of Business at The University of Calgary. His research interests focus on the labor market and employment. Among his publications and studies are *The Effects of Technological Changes on the Educational and Skill Requirements of Industry, Computer Technology and Employment: Retrospect and Prospect*, and *Issues in Labour-Management Relations in the 1990's.*

J. RICHARD PONTING is Associate Professor of Sociology at The University of Calgary, where he specializes in Canadian society, public policy, and minority relations. He has conducted national, provincial, and local public opinion surveys on various aspects of public policy, and is currently Associate Editor of *Canadian Public Policy* (Sociology/ Anthropology). He has conducted research on various aspects of organizational response to disasters, including one project which assessed the disaster preparedness of a large nuclear power generating facility.

LARRY PRATT is Professor in the Department of Political Science at the University of Alberta. He has served on the editorial board of *Studies in Political Economy; Canadian Journal of Political Science*. He was nominated for the Governor General's Award for Literature in 1980 and received the Certificate of Merit from the Canadian Historical Society as co-author of *Prairie Capitalism*. His other works include *The Tar Sands: Syncrude and the Politics of Oil*, *Western Separatism*, and articles in *Studies in Political Economy* and *International Affairs*.

WILLIAM REEVES is Associate Professor of Sociology at The University of Calgary. His previous publications, among them *Librarians as Professionals*, have focused upon the role of organizations in structuring and defining status groups.

ALLAN SCHNAIBERG is Professor and former Chairman of the Department of Sociology at Northwestern University. In recent years he has served as a consultant to the International Institute of Environment and Society at the Science Center-Berlin, and as a member of the National Academy of Science Committee on Behavioral and Social Aspects of Energy Consumption. He is the author of *The Environment: From Surplus to Scarcity*, and of recent papers on conflicts surrounding issues of appropriate technology and environmental-energy policy. His research interests have focused on social inequality factors in environmental problems and policies, including the systematic biases in information systems used to facilitate policy making.

ANDREW THOMPSON is Professor of Law and Director of the Westwater Research Centre at The University of British Columbia and specializes in natural resource subjects. He has been Chairman of the British Columbia Energy Commission and conducted a federal public inquiry into west coast oil port proposals. He is author of *Canadian Oil and Gas*, an eight-volume treatise covering oil and gas law and legislation in Canada, and has contributed numerous articles and publications in the natural resources and environmental fields.